Homework Helpers:
Algebra

HOMEWORK HELPERS

Algebra

By

Denise Szecsei

CAREER
PRESS

Pompton Plains, NJ

HOMEWORK HELPERS: ALGEBRA

TYPESET BY EILEEN MUNSON

Original cover design by Lucia Rossman, Digi Dog Design
Printed in the U.S.A.

To order this title, please call toll-free 1-800-CAREER-1 (NJ and Canada: 201-848-0310) to order using VISA or MasterCard, or for further information on books from Career Press.

CAREER
PRESS

The Career Press, Inc.
220 West Parkway, Unit 12
Pompton Plains, NJ 07444
www.careerpress.com

Library of Congress Cataloging-in-Publication Data
CIP data available upon request.

Acknowledgments

This book was a group effort, and I would like to thank the people who helped transform it from the electrons on my computer screen into the object you are holding in your hand.

I would like to thank Michael Pye, Kristen Parkes, and everyone else at *Career Press* who worked on this project. I appreciate the time and efforts of Jessica Faust, who was instrumental in making the connections that started things rolling.

Kendelyn Michaels played a pivotal role throughout the development of this book. I benefited greatly from her review of manuscript, and I have never met anyone who comes close to her level of thoroughness and consistency.

Alic Szecsei helped reduce the number of typographical errors in the manuscript and had the privilege of working out the problems in this book after he finished doing his *own* algebra homework.

Thanks to my family for their help throughout the writing and editing stage. The extra chores that they did and the dinners that they made did not go unnoticed. The only problems they gave me were the ones that I put in my book.

C
O
N
T
E
N
T
S

Welcome to Homework Helpers Algebra!

Algebra marks the transition from learning how to do basic calculations to being able to solve more complex and interesting problems. You have had enough practice adding, subtracting, multiplying, and dividing numbers, and are ready to solve more interesting problems. Knowing that the price of a video game is $50 is one thing, but being able to calculate how many months you will have to save your allowance in order to afford to buy the game is quite another! The more lofty your goals, the greater the role that algebra will play in helping you achieve them.

Algebra involves taking a step back from doing rote calculations and emphasizes looking at the big picture. Mathematics is all about discovering patterns and pushing the limits to develop new methods for solving problems, and algebra will give you a taste of more advanced problem-solving. Hopefully it will whet your appetite and leave you wanting to learn more.

Algebra has gotten a bad reputation over the years. Algebra seems to have turned into a subject that people "survive" rather than enjoy. Algebra "survivors" spread tales of the trials and tribulations that they suffered while getting through the material, and few people speak on its behalf. Let me speak up for algebra.

I don't have a particularly problematic life, yet I find that I use algebra to solve problems on a daily basis. And that's not just because I teach math! I enjoy using algebra to solve problems. I usually turn situations in which I use algebra into word problems. Word problems provide you (the student) with a glimpse of how algebra can be used to solve problems that occur in everyday life. They also provide a creative outlet for math teachers.

The skills that you will develop in the process of learning algebra will carry over into every other field of study imaginable. Philosophy, psychology, sociology, medicine, and *all* fields of science are just a few areas where your enhanced mathematical problem-solving skills will enable you to excel. Algebra is not just an area of mathematics. It is a way of life!

I wrote this book with the hope that it will help anyone who is struggling to understand algebra or needs to have their math skills refreshed. Reading a math book can be a challenge, but I tried to use everyday language to explain the concepts being discussed. Looking at solutions to algebra problems can sometimes be confusing, so I tried to explain each of the steps I used to get from Point A to Point B. Keep in mind that learning algebra is not a spectator sport. In this book, I have worked out many examples, and I have supplied practice problems at the end of most lessons. Work these problems out on your own as they come up, and check your answers against the solutions at the end of the book. Aside from any typographical errors on my part, our answers should match.

Perhaps you aren't quite convinced that algebra is all that I am making it out to be. Approach the subject with an open mind, take this book home and let your journey begin!

1

Numbers

Numbers are tools that you have been working with for many years. You have learned how to add, subtract, multiply, and divide. You can count how much money you have, and you can calculate how much more you will need to buy something you want, but can you figure out how long it will take you to make up the difference? Algebra is a way to use numbers to answer more advanced questions.

Algebra can be thought of as a language of numbers. Numbers are the tools used to communicate mathematical ideas. This chapter will focus on revisiting the rules that must be followed in order to use numbers effectively.

The topics in algebra will apply to *all* numbers. There are infinitely many numbers, and no one has the time to address each and every number individually. Instead, I will talk about numbers in general. One way to talk about numbers in general is to use a letter to represent *any* number. When we use a letter to represent any number, the letter is called a **variable**.

Lesson 1-1: Types of Numbers

The best numbers to start with are the **counting numbers**, which are the numbers we use to count. We start counting with the number 1, then move on to 2, 3, and 4. This collection of numbers is also called the **natural numbers**, or the **positive integers**. There is no end to the counting numbers. To represent that they continue forever, mathematically we write 1, 2, 3, 4, ... The three dots mean that the list of numbers is unending.

Zero was the last number to be discovered. It is a symbol used to denote "nothing." If you add zero to the collection of counting or natural numbers, you will have the whole numbers. The **whole numbers** are the numbers 0, 1, 2, 3, ...

Numbers can be used to count things that you have, and they can also count things that you owe. Negative integers are thought of as the *opposite* of the counting numbers. The **negative integers** are the numbers –1, –2, –3, –4, ...

The **integers** consist of the positive integers, the negative integers and zero. They can be written ..., –3. –2, –1, 0, 1, 2, 3, ... Notice that the dots represent that the numbers go on forever in both directions.

A **rational number** is a number that can be written as the ratio of two integers. For example, the numbers $\frac{1}{2}$, $-\frac{2}{3}$, and $\frac{10}{3}$ are rational numbers. A rational number is a number that can be written as $\frac{p}{q}$, where p and q are integers and q is any number other than zero. The number p is called the **numerator** and q is called the **denominator**. Notice that there's more than one way to represent a given rational number: $\frac{2}{3}$, $\frac{6}{9}$, and $\frac{-20}{-30}$ all represent the same rational number. The integer 2 is a rational number: $2 = \frac{2}{1}$. In fact, *every* integer is a rational number. Whenever you are working with whole numbers and fractions in the same problem, remember that a whole number can be thought of as a fraction whose denominator is 1.

All rational numbers also have a decimal representation that either terminates, or it repeats. The numbers $\frac{3}{4} = 0.75$ and $\frac{3}{5} = 0.6$ have decimal representations that terminate. The numbers $\frac{1}{3} = 0.3333...$, $\frac{5}{11} = 0.454545...$, and $-\frac{131}{333} = -0.393393...$ have decimal representations that do not terminate, but repeat a pattern. Rational numbers will be discussed in more detail in Lesson 1-6.

Not all decimal representations have to terminate or repeat. Numbers that are written as non-terminating and non-repeating decimals are called **irrational numbers**. An example of an irrational number is $\sqrt{2}$. An irrational number cannot be written as the ratio of two integers.

Taken together, the rational numbers and the irrational numbers form the set of **real numbers**. Breaking the real numbers into groups (the natural numbers, integers, rational numbers, irrational numbers, and real numbers), is one way to classify numbers in general.

Another way to classify a natural number is to look at the numbers that divide it evenly. When one number divides another number **evenly**, the remainder is 0. *Every* number is evenly divisible by 1, and *every* non-zero number is evenly divisible by itself. The number 1 is called a **trivial factor** (or divisor) because it divides evenly into every other number. The number 1 is the *only* trivial factor, because 1 is the only number that divides evenly into every other number. All other factors of a number are considered to be **non-trivial**.

A number is **even** if it is evenly divisible by 2. For example, 22 is even because when you divide 22 by 2 the remainder is 0. A number is **odd** if it is not evenly divisible by 2. For example 15 is odd because when you divide 15 by 2 the remainder is 1 (not 0). Determining whether a number is even or odd is one example of a classification of numbers based on the numbers that divide them.

A number is **prime** if the only numbers that divide into it evenly are 1 and itself. For example, 7 is prime because the only numbers that divide into it and leave a remainder of 0 are 1 and 7. A number is **composite** if there are numbers other than 1 and itself that divide into it evenly. The number 6 is composite because, besides 1 and 6, 2 divides into it evenly (as does 3). There is only one even number that is also a prime number: 2. All other even numbers are divisible by 2, and hence are composite. Odd numbers may be prime or composite. Examples of odd *prime* numbers are 3, 5, and 7; examples of odd *composite* numbers are 9 and 15. All natural numbers *greater than 1* can be classified as either prime or composite.

Besides looking at individual numbers on their own, we can also look at pairs of numbers. One way to classify a pair of numbers is according to the numbers that divide each number in the pair. A pair of numbers is **relatively prime** if the only common factor of the two numbers is 1. The numbers 15 and 8 are an example of a pair of numbers that are relatively prime. The only non-trivial factors of 15 are 3, 5, and 15, and the only non-trivial factors of 8 are 2, 4 and 8, so the only number that divides both 15 and 8 is 1. Notice that both 15 and 8 are *composite* numbers, when classified individually, yet they are *relatively prime* when classified as a pair. The numbers 7 and 21 are *not* relatively prime, because 7 is a non-trivial factor of both 7 and 21. So even though 7 is a prime number, it is *not* relatively prime to 21.

The **greatest common divisor**, or the **greatest common factor** of a pair of numbers is the largest number that evenly divides into both numbers. If two numbers are relatively prime then their greatest common factor is 1. Actually, the only common factor of two relatively prime numbers is 1. If two numbers are not relatively prime then their greatest common factor will be greater than 1. One way to find the greatest common factor of a pair of numbers is to first factor each number into its prime factors. Then match up, prime by prime, the prime numbers that appear in both factorizations. If any prime numbers repeat, you must take that into consideration as we will see in the second example.

Example 1

Find the greatest common factor of 12 and 15.

Solution: Since $12 = 2 \times 2 \times 3$ and $15 = 3 \times 5$, 3 is the largest number that evenly divides 12 and 15.

Thus the greatest common factor of 12 and 15 is 3.

Example 2

Find the greatest common factor of 60 and 36.

Solution: Factoring both numbers we have $60 = 2 \times 2 \times 3 \times 5$ and $36 = 2 \times 2 \times 3 \times 3$. Both 60 and 36 have a factor of 3 that appears once and a factor of 2 that appears twice.

The greatest common factor is $2 \times 2 \times 3$ or 12.

Given any two numbers you can always find a number that is evenly divisible by both of them. Any multiple of the product of the two numbers would work. The **least common multiple** of a pair of numbers is the smallest number that both numbers divide evenly into. One way to find the least common multiple of a pair of numbers is to factor both numbers into their prime factors and look at the list. For each prime factor, note the greatest number of times that it appears in either of the two factor trees. Use this information to construct a product of prime factors that will generate the least common multiple. I will walk you through the process in the next example.

Example 3

Find the least common multiple of 45 and 24.

Solution: Factor 45 and 24:

$45 = 3 \times 3 \times 5$

$24 = 2 \times 2 \times 2 \times 3$

Write down each prime factor and the greatest number of times it appears in either of the two factor trees:

1

NUMBERS

Prime factor	Greatest number of times it appears
2	3
3	2
5	1

Now find the least common multiple:

$2 \times 2 \times 2 \times 3 \times 3 \times 5 = 360$

The least common multiple of 45 and 24 is 360.

Another way to approach this problem involves finding the greatest common factor of the two numbers. Multiply the original two numbers together and then divide by the greatest common factor. This result will be the least common multiple.

Example 4

Find the least common multiple of 12 and 15.

Solution: As discussed in Example 1, the greatest common factor of 12 and 15 is 3. The least common multiple of 12 and 15 is found by multiplying 12×15 and then dividing by 3:

$(12 \times 15) \div 3 = 180 \div 3 = 60$

The smallest number that both 12 and 15 divide evenly into is 60.

We will be using the greatest common factor and the least common multiple throughout this book. Work the following problems before moving on to the next lesson. The answers are given at the end of the chapter.

Lesson 1-1 Practice

Address the following.

1. Determine whether the following 8 statements are true or false.

    ☐ True   ☐ False   a. –4 is an integer.

    ☐ True   ☐ False   b. –3 is a natural number.

    ☐ True   ☐ False   c. 5 is a rational number.

□ True　□ False　d. 0 is not a rational number.

□ True　□ False　e. $\sqrt{2}$ is a rational number.

□ True　□ False　f. $\frac{7}{0}$ is a rational number.

□ True　□ False　g. $\frac{0}{7}$ is a rational number.

□ True　□ False　h. Every integer is either positive or negative.

2. Find the greatest common factor and the least common multiple of 20 and 35.

Lesson 1-2: Operations and Symbols

Success in algebra hinges on your ability to manipulate real numbers. There are many things that you can do with numbers. The top four operations that come to mind are adding, subtracting, multiplying and dividing. While there is just one symbol for addition, and one symbol for subtraction, there are several symbols for multiplication and

Symbol	Meaning	Example
+	Addition	$3 + 2 = 5$
−	Subtraction	$3 - 2 = 1$
$\times$, $\cdot$ or $(\ )(\ )$	Multiplication	$3 \times 2 = 6; 3 \cdot 2 = 6;$ $(3)(2) = 6$
$\div, -,$ or $/$	Divided by	$3 \div 2 = \frac{3}{2} = \frac{3}{2} = 1\frac{1}{2}$
$=$	Is equal to	$2 = 2$
$\neq$	Is not equal to	$3 \neq 2$
$>$	Is greater than	$3 > 2$
$\geq$	Is greater than or equal to	$3 \geq 3, 3 \geq 2$
$<$	Is less than	$2 < 3$
$\leq$	Is less than or equal to	$3 \leq 3, 2 \leq 3$

division. Besides manipulating numbers, you can also compare them. To do that, you will need symbols for "less than" and "greater than." I'll list the main symbols that we will use here, and then give some examples of how they are used.

The final concept to be discussed in this section has to do with describing whether a number is positive, negative, or zero. If a number is **positive** then we mean that it is greater than 0. For example, 2 is a positive number, and we can write $2 > 0$. If a number is **negative** then we mean that it is less than 0. For example, –3 is a negative number, and we can write $-3 < 0$. You'll soon discover that comparing numbers to 0 can be very useful.

Lesson 1-3: Properties of Real Numbers

There are several properties that all real numbers share. Because these are properties that *all* real numbers have, I will use the variables a, b and c to represent *any* real number.

- ▫ **Closure:** The first property that all real numbers have is **closure** under addition and multiplication. In other words, the set of real numbers has the property that if you take any two real numbers and add them (or multiply them), what you'll end up with is a real number. It doesn't have to be the same number, though it may be. The important point is that if you combine any two real numbers by addition or multiplication you will always get a real number. We say that the set of real numbers is **closed** under addition and multiplication.

- ▫ **Commutative property of addition and multiplication:** The order in which two numbers are added or multiplied doesn't matter. For example, $2 + 3 = 3 + 2$, and $(8)(-3) = (-3)(8)$. In other words, addition and multiplication **commute**. This is written symbolically as $a + b = b + a$ and $a \times b = b \times a$.

- **Associative property of addition and multiplication:** If you have a long list of numbers to add (or multiply), then you can group them in any order.

    For example, $1 + (2 + 3) = (1 + 2) + 3$, and $2 \cdot (\frac{1}{2} \cdot 3) = (2 \cdot \frac{1}{2}) \cdot 3$. In other words, addition and multiplication are *associative*. This is written symbolically as $a + (b + c) = (a + b) + c$ and $a \times (b \times c) = (a \times b) \times c$.

The next two properties have more to do with the relationships between numbers.

- **Trichotomy property of real numbers:** When you compare two numbers a and b, only one of three things can be true:

    1. $a < b$
    2. $a = b$
    3. $a > b$

    This is known as the **trichotomy property** of real numbers. In fact, you can compare any two numbers using these three relations.

- **Transitive property of equality:** It two numbers are both equal to a third number, then the two numbers are equal to each other. This property is called the **transitive property of equality**. It can be stated more generally in terms of variables: if $a = b$ and $b = c$, then $a = c$. For example, if $5 + 5 = 10$ and $10 = 2 \times 5$, so $5 + 5 = 2 \times 5$.

At first glance it may appear that all numbers are created equally, but there are actually two numbers that deserve special attention. Those two numbers are 0 and 1. Zero is the *only* number that you can add to any other number and have no effect. Zero is called the **additive identity**, and this property can be written in general using the equation $a + 0 = a$; for example, $2 + 0 = 2$. It may not seem very important...after all, nothing is nothing. But quantifying nothing is not as trivial as you might think.

The number 1 is important for a similar reason; 1 is the **multiplicative identity**. It is the *only* number that you can multiply any other number by and have no effect. This idea is written in general using the equation $a \times 1 = a$; for example, $3 \times 1 = 3$. The numbers 0 and 1 will play a significant role in solving many algebra problems.

The additive identity plays a role in the development of subtraction. It turns out that for every real number a, there is a unique real number, called the **additive inverse** of a, and denoted $-a$, such that $a + (-a) = 0$. In other words, the additive inverse of a, or negative a, is the unique number that you *add* to a to get 0 (the additive identity). Every real number has an additive inverse. Notice that 0 is its own additive inverse. Subtraction is then defined in terms of addition: $a - b = a + (-b)$. We sometimes refer to the additive inverse of a number as the opposite of the number. For example, the opposite of 2 is -2, and the opposite of -2 is 2. Opposite, negation, and additive inverse all mean the same thing and are used interchangeably.

You may have been told that a negative number times a negative number is a positive number. The reason for that stems from the fact that $-a$ is the opposite of a: $-a$ is the unique number that, when it is added to a, gives you 0. In other words, $a + (-a) = 0$. What would be the opposite of $-a$? Well, in keeping with our notation, it would be $-(-a)$. But wait a minute! The opposite of $-a$ is a. Now it appears that you have two opposites of $-a$: a and $-(-a)$, but you can't have two different opposites of a number. Additive inverses are unique, which means that each number can only have one additive inverse. The only way for this to make sense is if a and $-(-a)$ are the same thing: $-(-a) = a$. We can apply this result specifically to the number 1: $-(-1) = 1$. This is interpreted as meaning that a negative times a negative equals a positive.

It's time to turn our attention to multiplicative inverses. For any real number a (except 0), there is a unique real number, called the **multiplicative inverse** and denoted a^{-1}, satisfying the equation $a \cdot a^{-1} = 1$. For example, since $2 \times \frac{1}{2} = 1$, the multiplicative inverse of 2 is $\frac{1}{2}$, and the multiplicative inverse of $\frac{1}{2}$ is 2. The multiplicative inverse of a is the unique number that you *multiply* a by to get 1 (the multiplicative identity). Notice that 0 (the additive inverse) is the *only* real number that doesn't have a multiplicative inverse. The multiplicative inverse of a number is also called the **reciprocal** of that number.

There are two common ways to represent the multiplicative inverse of a; it can be written as a^{-1} or as $\frac{1}{a}$.

Division can then be defined in terms of multiplication.

$$\text{If } b \neq 0, \text{ then } a \div b = \frac{a}{b} = a\left(\frac{1}{b}\right) = a\left(b^{-1}\right)$$

The reciprocal of $\frac{a}{b}$ is just $\frac{b}{a}$.

The last property of the real numbers that I will discuss in this section has to do with how to combine addition and multiplication. It is called the **distributive property**. The distributive property states that multiplication **distributes** over addition and can be expressed using the formula:

$$a \times (b + c) = a \times b + a \times c$$

You can think of a as being distributed to both b and c. I will have more to say about the distributive property in Lesson 1-8.

There are several properties of real numbers that you are probably familiar with. I've summarized some of the most useful ones here, and have included some examples shown on page 24 to help illustrate the ideas involved.

Lesson 1-4: Properties of Zero and One

As I have already mentioned, 0 and 1 are very special numbers. Being familiar with their properties will help you solve many algebra problems.

Property	Example
$-a = -1 \times a$	$-3 = -1 \times 3$
$-(-a) = a$	$-(-3) = 3$
$a - b = a + (-b)$	$3 - 2 = 3 + (-2)$
$a - (-b) = a + b$	$3 - (-4) = 3 + 4$
$(-a) \times b = -a \times b = a \times (-b)$	$(-3) \times 2 = -3 \times 2 = 3 \times (-2)$
$a \times (b - c) = a \times b - a \times c$	$3 \times (6 - 2) = 3 \times 6 - 3 \times 2$
$-(a + b) = -a - b$	$-(3 + 6) = -3 - 6$
$-(a - b) = -a + b$	$-(5 - 3) = -5 + 3$
$(-a)(-b) = a \times b$	$(-2)(-3) = 2 \times 3$
$0 \times a = 0$	$0 \times 3 = 0$
$\dfrac{a}{1} = a$	$\dfrac{6}{1} = 6$
$\dfrac{0}{a} = 0 \quad (a \neq 0)$	$\dfrac{0}{4} = 0$
$\dfrac{a}{a} = 1 \quad (a \neq 0)$	$\dfrac{3}{3} = 1$
$a\left(\dfrac{b}{a}\right) = b \quad (a \neq 0)$	$3\left(\dfrac{2}{3}\right) = 2$
$\dfrac{a}{-b} = -\dfrac{a}{b} = \dfrac{-a}{b}$	$\dfrac{2}{-7} = -\dfrac{2}{7} = \dfrac{-2}{7}$
$\dfrac{-a}{-b} = \dfrac{a}{b}$	$\dfrac{-3}{-8} = \dfrac{3}{8}$

NUMBERS

1

There are two important characteristics of 0: 0 is its own additive inverse, and 0 is the only real number that doesn't have a reciprocal (or multiplicative inverse). It is worth exploring why 0 doesn't have a reciprocal. Whenever you multiply 0 and any number a, the result is always 0: $0 \times a = 0$. Remember that the reciprocal of a number is the number you multiply by in order to get 1. If 0 had a reciprocal, say b, then that would mean that the product of 0 and b would be 1! But that can't happen, because the product of 0 and any number is 0. So 0 is stuck. It is precisely because 0 times any number is 0 that 0 can't have a reciprocal. This fact translates into the idea that $\frac{1}{0}$ (which means the reciprocal of 0) is meaningless.

The other special property of 0 follows along a similar line of thought. If two numbers are multiplied together and the result is 0, then one thing is certain: at least one of the two numbers has to be 0. To write this idea using equations, if a and b are any two numbers satisfying $a \times b = 0$, then either $a = 0$ or $b = 0$. It is possible that both a and b are zero, since 0×0 is certainly 0. The key idea is that when you have to compare a product of numbers to something, the best number to compare the product to is 0! No other real number has this property. If two numbers multiply together to equal 2, there is nothing you can say about the two numbers. If $a \times b = 2$, then a and b can take on any value that they want, as long as their product is 2. There are lots of pairs of numbers whose product is 2: $1 \times 2 = 2$, $4 \times \frac{1}{2} = 2$, and $6 \times \frac{1}{3} = 2$ are just a few. I suppose there is one thing you can say for sure about two numbers whose product is 2: neither of the two numbers are 0! This multiplication property is very important, and 0 is the *only* real number that has it.

I've said enough about 0. It's time to turn our attention to 1. The main properties we will make use of now are $\frac{a}{a} = 1$, and $a\left(\frac{b}{a}\right) = b$. We can combine these to get another useful equation:

$$\left(\frac{a}{b}\right)\left(\frac{b}{a}\right) = 1$$

1

NUMBERS

This just means that the reciprocal of $\frac{a}{b}$ is $\frac{b}{a}$. We will make use of these properties when we reduce fractions in Lesson 1-6. And the importance of the number 1 will be clear when you start solving algebraic equations.

Lesson 1-5: Absolute Value

The **absolute value** of a real number represents the size, or *magnitude*, of that number. It can be interpreted as how far away from 0 the number is on the number line. The absolute value of a number a is always positive, and is written symbolically as $|a|$. The absolute value of a positive number is itself. We can write this symbolically as:

$$\text{If } a > 0, \text{ then } |a| = a$$

The absolute value of 0 is 0, and the absolute value of a negative number is the opposite of that number. In other words, since –3 is 3 units away from 0 we have that $|-3| = 3$. We can write in general as:

$$\text{If } a < 0, \text{ then } |a| = -a$$

For example, $|8| = 8, |0| = 0$, and $|-5| = -(-5) = 5$. Notice that $|8| = 8$ and $|-8| = -(-8) = 8$. In other words, $|8| = |-8|$. This idea holds, not just for the number 8, but for all real numbers: if a is any real number, then $|a| = |-a|$. We can write the absolute value of a number using the formula:

$$|a| = \begin{cases} -a & \text{if } a < 0 \\ a & \text{if } a \geq 0 \end{cases}$$

You can read this formula as a fork in the road. The absolute value of a number depends on the sign of the number inside the absolute value symbols: $|a| = -a$ if $a < 0$ and $|a| = a$ if $a \geq 0$. In order to use this formula you must first check to see whether the contents of the absolute value symbols are positive or negative.

Example 1

Find the following:

a. $|-4|$

b. $|2|$

Solution:

a. In order to find $|-4|$, first look at the number inside of the absolute value symbol: the number is –4. This is a negative number (–4<0) so we need to use the top rule: $|-4|=-(-4)=4$.

b. In order to find $|2|$, first look at the number inside of the absolute value symbol: the number is 2. This is a positive number (2>0) so we need to use the bottom rule: $|2|=2$.

Lesson 1-6: Manipulating Rational Numbers

Rational numbers are sometimes called **simple fractions**. They are formed by dividing one integer by another integer. Remember from Lesson 1-1 that we can represent a rational number by: $\frac{a}{b}$, where both a and b are integers and $b \neq 0$. We call a the numerator and b the denominator. We also write the rational number $\frac{a}{b}$ using the ÷ symbol: $a \div b$. If both a and b are positive integers and $a < b$, then $\frac{a}{b}$ is called a **proper fraction**. If both a and b are positive integers and $a > b$ then $\frac{a}{b}$ is called an **improper fraction**. If both a and b are positive integers and b divides evenly into a, then $\frac{a}{b}$ is a positive integer.

Improper fractions can be written as **mixed numbers**, which are numbers with an integer part and a proper fraction part. For example, the improper fraction $\frac{5}{2}$ can be written as the mixed number $2\frac{1}{2}$.

Multiplying Fractions

The rules for multiplying fractions are very straightforward. The product of two fractions is found by multiplying the two numerators together to get a new numerator, and multiplying the two denominators together to get a new denominator. The new numerator and the new denominator make up your new fraction.

Example 1

Perform the following multiplication: $\dfrac{2}{5} \times \dfrac{7}{11}$

Solution: $\dfrac{2}{5} \times \dfrac{7}{11} = \dfrac{14}{55}$

Reducing fractions

Most people find it easier to work with smaller numbers than larger numbers. If the numerator and denominator of a rational number $\dfrac{a}{b}$ have common factors, it's best to put this rational number into reduced form. A rational number is in **reduced form** if the numerator and the denominator are relatively prime. In order to write a rational number in reduced form, you must completely factor both a and b, and then use the properties $\dfrac{a}{a} = 1$, $a\left(\dfrac{b}{a}\right) = b$ and $1 \times a = a$.

Example 2

Write the fraction $\dfrac{28}{48}$ in reduced form.

Solution: The first step in reducing a fraction is to find the greatest common divisor of 28 and 48. Completely factor 28 and 48:

$28 = 2 \times 2 \times 7$

$48 = 2 \times 2 \times 2 \times 2 \times 3$

Both 28 and 48 are evenly divisible by 2×2, or 4, so we can write:

$$\dfrac{28}{48} = \dfrac{4 \times 7}{4 \times 12} = \dfrac{4}{4} \times \dfrac{7}{12} = 1 \times \dfrac{7}{12} = \dfrac{7}{12}$$

As you become familiar with the process, you will find yourself leaving out some of the intermediate steps and writing something like:

$$\dfrac{28}{48} = \dfrac{\cancel{2} \times \cancel{2} \times 7}{\cancel{2} \times \cancel{2} \times 2 \times 2 \times 3} = \dfrac{7}{12}, \quad \text{or}$$

$$\dfrac{28}{48} = \dfrac{\cancel{4} \times 7}{\cancel{4} \times 12} = \dfrac{7}{12}$$

This process is often referred to as *canceling the 4s*. What you are really doing is making use of those useful properties of multiplication I mentioned earlier. Remember that the key to reducing fractions is to find the greatest common denominator by factoring the numerator and the denominator completely. You can only cancel terms that are being *multiplied* together.

It is your turn to practice what we have discussed. You should work these problems before moving on. Be sure to check your answers.

Lesson 1-6a Practice

Write the following fractions in reduced form:

1. $\dfrac{15}{33}$

2. $\dfrac{27}{30}$

3. $\dfrac{21}{77}$

Multiplying Fractions and Reducing

When you multiply two fractions together, the numbers can get awfully big. And, after you multiply the two fractions together, you must write the resulting fraction in reduced form. You may find yourself factoring big numbers in the process, but there is a better way. Instead of multiplying numerators together and denominators together, factor both numerators and both denominators and try to cancel out common factors *before* multiplying them. That will make the process much easier.

Example 3

Find the product: $\dfrac{5}{12} \times \dfrac{63}{130}$

Write your answer in reduced form.

Solution: Rather than finding the products 5 × 63 and 12 × 130, it's better to factor the numerators and denominators completely first and then cancel what you can:

$$\frac{5}{18} \times \frac{63}{130} = \frac{\cancel{5} \times 7 \times \cancel{3} \times \cancel{3}}{2 \times \cancel{3} \times \cancel{3} \times 13 \times 2 \times \cancel{5}} = \frac{7}{52}$$

Try these problems, just to make sure you have the techniques down.

Lesson 1-6b Practice

Find the following products. Write your answers in reduced form.

1. $\dfrac{6}{35} \times \dfrac{20}{27}$ 2. $\dfrac{11}{15} \times \dfrac{21}{22}$ 3. $\dfrac{7}{36} \times \dfrac{66}{85}$

Dividing Fractions

As a teacher, I find that students have a hard time understanding why anyone would ever need to divide a number by a fraction. Dividing an integer by another integer is probably more common, but dividing by a fraction is useful in solving problems like determining how many arcade games you can play $10 worth of quarters. Understanding *why* you would want to divide a number by a fraction will hopefully motivate you to master the technique involved.

Remember that a simple fraction is a ratio of two integers. A **complex fraction** is a fraction where the numerator, denominator, or both, are fractions. A complex fraction can be converted to a simple fraction by multiplying the numerator by the reciprocal of the denominator:

$$\frac{\frac{a}{b}}{\frac{c}{d}} = \frac{a}{b} \times \frac{d}{c}$$

The reason for this is because of the properties of multiplication mentioned earlier.

Start with the expression: $\dfrac{\frac{a}{b}}{\frac{c}{d}}$

Multiply by 1 disguised as the ratio of the denominator to the denominator: $\dfrac{\frac{c}{d}}{\frac{c}{d}}$

$$\frac{\frac{a}{b}}{\frac{c}{d}} = 1 \times \frac{\frac{a}{b}}{\frac{c}{d}} = \frac{\frac{d}{c}}{\frac{d}{c}} \times \frac{\frac{a}{b}}{\frac{c}{d}}$$

This is starting to look pretty complicated, but now we are just multiplying two fractions. We will multiply the numerators together and the denominators together. Notice that the product in the denominator is $\frac{d}{c} \times \frac{c}{d}$, which is equal to 1:

$$\frac{\frac{d}{c} \times \frac{a}{b}}{\frac{d}{c} \times \frac{c}{d}} = \frac{\frac{d}{c} \times \frac{a}{b}}{1} = \frac{a}{b} \times \frac{d}{c}$$

The expression $\frac{\frac{a}{b}}{\frac{c}{d}}$ is equivalent to the expression $\frac{a}{b} \times \frac{d}{c}$, which is the product of the numerator and the inverted denominator. The net result is that when you divide one fraction by another fraction you invert the fraction in the denominator and multiply. This is usually shortened to just "invert and multiply" but you have to remember that it is the denominator that gets inverted!

Example 4

Simplify the complex fraction: $\dfrac{\frac{3}{4}}{\frac{2}{5}}$

Solution:

Invert the denominator and multiply: $\dfrac{\frac{3}{4}}{\frac{2}{5}} = \frac{3}{4} \times \frac{5}{2} = \frac{15}{8}$

You could also write your answer as a mixed number: $\dfrac{15}{8} = 1\dfrac{7}{8}$

Here is yet another opportunity to try your hand at dividing fractions. Be sure to check your answers before moving on to the next topic.

Lesson 1-6c Practice

Simplify the following complex fractions. Write your answer in reduced form.

1. $\dfrac{\frac{3}{7}}{\frac{4}{9}}$

2. $\dfrac{\frac{4}{9}}{\frac{10}{27}}$

3. $\dfrac{\frac{7}{15}}{\frac{3}{10}}$

Adding and Subtracting Fractions

It may seem strange that our discussion on fractions started with multiplication and division before addition and subtraction, especially since multiplication can be interpreted as repetitive addition. With fractions, addition and subtraction is more complicated than multiplication and division. In fact, adding and subtracting fractions can actually involve multiplying fractions as part of the overall process.

Two fractions can be added or subtracted *only* if they have the same denominator: all you do is add or subtract the numerators and keep the denominator the same. Once you have added or subtracted the numerators, you must try to reduce the resulting fraction by looking for factors that are common to the resulting numerator and the denominator.

Example 5

Simplify: $\dfrac{8}{15} - \dfrac{2}{15}$ Write your answer in reduced form.

Solution: Since the denominators are the same, subtract one numerator from the other and try to reduce:

$$\frac{8}{15} - \frac{2}{15} = \frac{8-2}{15} = \frac{6}{15} = \frac{2 \times \cancel{3}}{\cancel{3} \times 5} = \frac{2}{5}$$

When the denominators are different, adding and subtracting fractions can get a bit tricky. Remember that the only time you are allowed to add or subtract two fractions is if their denominators are the *same*. If the two fractions have different denominators you have to turn them into fractions that have the same denominator. The only tool you have at your disposal is to multiply by 1. But you are allowed to use a more complicated form of 1.

Suppose you want to add the fractions $\dfrac{1}{3}$ and $\dfrac{1}{2}$. Clearly these two fractions have different denominators, so you can't combine the two fractions. You must change these fractions so that they have the same denominator, and the best denominator to use is the least common multiple of the two denominators; the least common multiple of 3 and 2 is 6. We will convert each fraction into a form that has 6 as the denominator by multiplying by 1 disguised as a fraction:

$$\frac{1}{3} = 1 \times \frac{1}{3} = \frac{2}{2} \times \frac{1}{3} = \frac{2}{6}$$

$$\frac{1}{2} = 1 \times \frac{1}{2} = \frac{3}{3} \times \frac{1}{2} = \frac{3}{6}$$

Now that we have written $\frac{1}{3}$ and $\frac{1}{2}$ as fractions with the same denominator we can add them together:

$$\frac{1}{3} + \frac{1}{2} = \frac{2}{6} + \frac{3}{6} = \frac{5}{6}$$

Once we've added the two fractions we must check whether or not we can reduce the resulting fraction. In this case, 5 and 6 are relatively prime so there is no further reduction.

Example 6

Find: $\dfrac{3}{25} + \dfrac{9}{20}$

Solution: In order to add these two fractions we need them to have a common denominator. First, factor both denominators to find the least common multiple: $25 = 5 \times 5$ and $20 = 2 \times 2 \times 5$ so the least common multiple is $2 \times 2 \times 5 \times 5 = 100$. Change each fraction so that its denominator is 100:

$$\frac{3}{25} = 1 \times \frac{3}{25} = \frac{4}{4} \times \frac{3}{25} = \frac{12}{100}$$

$$\frac{9}{20} = 1 \times \frac{9}{20} = \frac{5}{5} \times \frac{9}{20} = \frac{45}{100}$$

Finally, add the two fractions together:

$$\frac{3}{25} + \frac{9}{20} = \frac{12}{100} + \frac{45}{100} = \frac{57}{100}$$

Check to see if any reduction is possible. Because 57 and 100 are relatively prime, no further reduction is possible and the problem is solved.

Now it is your turn to apply the techniques discussed.

Lesson 1-6d Practice

Find the following sums.

1. $\dfrac{5}{62} + \dfrac{19}{62}$ 2. $\dfrac{5}{32} + \dfrac{15}{20}$ 3. $\dfrac{21}{55} + \dfrac{13}{20}$

Lesson 1-7: The Order of Operations

So far I have talked about four basic operations: addition, subtraction, multiplication, and division. When you see an expression like $3 + 4 \times 2 \div 3 - 1$, the order in which you perform the calculation matters. Some people may be tempted to read the expression from left to right, performing each operation as it is written. Others want to do their favorite operations first, and leave their least favorites until the end. The answer you get will vary depending on the order in which you tackle this expression. Because of this possible confusion (and many different "correct" results for the same expression), rules had to be established so that everyone will approach this problem in the same way and get the same answer. These rules are known as the **order of operations.**

The order of operations dictates that multiplication and division, read left to right, take precedence over addition and subtraction. For example, the expression $3 + 2 \times 5$ should evaluate to 13. You must first multiply 2 and 5, and then add 3. If you wanted to do the addition first and then multiply, you would need to use parentheses: $(3 + 2) \times 5$. The result of this calculation is 25. Keep in mind that when you want things to be done in an order that differs from the accepted order of operations you must use parentheses. Any operations inside parentheses must be done first, always following the standard order of operations. Once you have finished working out any expressions in parentheses you can turn your attention to the rest of the expression.

There are times when parentheses are assumed but not explicitly written. For example, the fraction $\dfrac{3+8}{3+19}$ should be treated as $\dfrac{(3+8)}{(3+19)}$

or $(3+8) \div (3+19)$. In order to simplify $\frac{3+8}{3+19}$ you must first do the addition and then try to reduce the resulting fraction:

$$\frac{3+8}{3+19} = \frac{11}{22} = \frac{1 \times \cancel{11}}{2 \times \cancel{11}} = \frac{1}{2}$$

Whenever your fraction involves addition or subtraction in either the numerator or the denominator you must first perform the addition or subtraction. You should practice visualizing the invisible parentheses. Remember that you cannot reduce a fraction unless the numerator and denominator are factored; you can only cancel common factors. Some people are tempted to cancel the 3's in a fraction like $\frac{3+8}{3+19}$. Let me emphasize that you are **never** allowed to cancel across an addition sign (not even on your birthday or in an emergency). Cancel only when there is multiplication involved.

Many people use a calculator to perform routine calculations. If you use a calculator it is important not only that you know the order of operations, but that you use a calculator that also knows the order of operations. Not all calculators are created equally; there are some calculators that do not know the order of operations. It's always a good idea to test drive a calculator, just to be sure that you are approaching problems the same way and are using the same order of operations.

You should practice a few calculations using the order of operations. You can also use a calculator to test whether your calculator knows the order of operations as well.

Lesson 1-7 Practice

Evaluate the following:

1. $3 + 4 \times 5 - 9$

2. $(4 + 9) \times 3$

3. $\dfrac{2+3}{2+23}$

Lesson 1-8: The Distributive Property

In Lesson 1-3 I mentioned one version of the distributive property: $a \times (b + c) = a \times b + a \times c$. The distributive property doesn't require the multiplication to be written on the left of the parentheses, multiplication distributes on either side. We could have written the distributive property as $(b + c) \times a = b \times a + c \times a$. The distributive property enables you to calculate products of numbers in two ways, and you can use whichever form is easier. Calculations that involve numbers can be evaluated using either the order of operations or the distributive property. There are times when evaluating an expression is easier using the order of operations, and there are other times when using the distributive property works to your advantage. I will give you a couple of examples to help you see the difference in the process.

It is easier to evaluate the expression $3 \times (2 + 8)$ using the order of operations than to use the distributive property:

$$3 \times (2 + 8) = 3 \times 10 = 30$$

The numbers are small and work out nicely. However, it is easier to evaluate the expression 11×52 using the distributive property:

$$11 \times 52 = 11 \times (50 + 2) = 550 + 22 = 572$$

Of course, when you need to expand expressions that involve both numbers and variables, the only option at your disposal is the distributive property. It is such a useful property that it is worth working out several examples.

Example 1

Use the distributive property to expand the following expressions:

a. $3(a + 4)$

b. $-(3 + a)$

c. $(-5) \times (3 - a)$

d. $(2 + 3a) \times 5$

e. $(5 - 2a) \times (-4)$

Solution:

a. $3(a + 4) = 3 \times a + 3 \times 4 = 3a + 12$

b. $-(3 + a) = (-1) \times (3 + a) = (-1) \times 3 + (-1) \times a = -3 - a$

c. $(-5) \times (3 - a) = (-5) \times 3 - (-5) \times a = -15 + 5a$

d. $(2 + 3a) \times 5 = 2 \times 5 + (3a) \times 5 = 10 + 15a$

e. $(5 - 2a) \times (-4) = 5 \times (-4) - (2a) \times (-4) = -20 + 8a$

We tend to use a calculator to perform routine calculations that involve only numbers, but being familiar with the multiplication process will make your algebraic life easier. Algebra is a generalization of these routine calculations. If you always rely on a calculator to work things out, you will have a harder time making the transition from calculations involving numbers to calculations involving variables. Here are some practice problems that only involve variables. You should work these problems out before going on to the next section.

Lesson 1-8 Practice

Use the distributive property to expand the following expressions:

1. $2(3a - 1)$

2. $(-9) \times (4a + 3)$

3. $(a + 8) \times (-3)$

4. $(5 - 2a) \times 4$

Lesson 1-9: Evaluating Expressions

An **algebraic expression** is a statement that combines numbers and variables together using any of the four operations you have learned about: addition, subtraction, multiplication and division. For example, the expression $a + 2$ just means that you take the number a (whatever that is) and add 2 to it. If someone told you that $a = 5$, then you would know that $a + 2$ would be 7 (which is $5+2$). In order to evaluate an expression for a particular value of the variables, just replace the

NUMBERS

1

variables with their particular values and then perform the calculation. Be sure to pay attention to the order of operations when you are performing the calculation.

Example 1

Evaluate the following algebraic expressions when $a = -2$ and $b = 3$:

1. $\dfrac{-3a}{a+b}$

2. $\dfrac{a-b}{a+b}$

3. $(2a + b)(b - a)$

Solutions:

1. $\dfrac{-3a}{a+b} = \dfrac{(-3)(-2)}{-2+3} = \dfrac{6}{1} = 6$

2. $\dfrac{a-b}{a+b} = \dfrac{-2-3}{-2+3} = \dfrac{-5}{1} = -5$

3. $(2a+b)(b-a) = (2\times(-2)+3)(3-(-2)) = (-4+3)(3+2)$
$$= (-1)(5) = -5$$

Here is your turn to practice evaluating algebraic expressions. Be sure to check your answers after you work these problems.

Lesson 1-9 Practice

Evaluate the following algebraic expressions when $a = 5$ and $b = -4$:

1. $\dfrac{-4b}{a+2b}$

2. $\dfrac{a+b}{2a-b}$

3. $(2a - 3b)(a + 2b)$

Answer Key
Lesson 1-1

1. a. True

 b. False; natural numbers are positive numbers.

 c. True; 5 can be written as $\dfrac{5}{1}$.

 d. False; 0 can be written as $\dfrac{0}{1}$.

 e. False; 2 is not a perfect square.

 f. False; you cannot divide by 0.

 g. True

 h. False; an integer can also be 0.

2. The greatest common factor of 20 and 35 is 5: $20 = 2 \cdot 2 \cdot 5$ and $35 = 5 \cdot 7$.

 The least common multiple of 20 and 35 is 140: $\dfrac{20 \cdot 35}{5} = 140$.

Lesson 1-6a

1. $\dfrac{5}{11}$ 2. $\dfrac{9}{10}$ 3. $\dfrac{3}{11}$

Lesson 1-6b

1. $\dfrac{8}{63}$ 2. $\dfrac{7}{10}$ 3. $\dfrac{77}{510}$

Lesson 1-6c

1. $\dfrac{27}{28}$ 2. $\dfrac{6}{5}$ or $1\dfrac{1}{5}$ 3. $\dfrac{14}{9}$ or $1\dfrac{5}{9}$

Lesson 1-6d

1. $\dfrac{12}{31}$

2. $\dfrac{29}{32}$: $\dfrac{5}{32} + \dfrac{15}{20} = \dfrac{25}{160} + \dfrac{120}{160} = \dfrac{145}{160} = \dfrac{29}{32}$

3. $\dfrac{227}{220}$ or $1\dfrac{7}{220}$: $\dfrac{21}{55} + \dfrac{13}{20} = \dfrac{84}{220} + \dfrac{143}{220} = \dfrac{227}{220} = 1\dfrac{7}{220}$

Lesson 1-7

1. 14

2. 39

3. $\dfrac{1}{5}$

Lesson 1-8

1. $6a - 2$

2. $-36a - 27$

3. $-3a - 24$

4. $20 - 8a$

Lesson 1-9

1. $-\dfrac{16}{3}$ or $-5\dfrac{1}{3}$

2. $\dfrac{1}{14}$

3. -66

2

Exponents

Exponents are shorthand for representing how many times a number is multiplied by itself. They are useful, in part, due to the fact that they can be used to represent numbers that are either very large, like the number of grains of sand on the beach, or very small in magnitude, like the mass of an atom. In this chapter we will discuss the properties of exponents and introduce the idea of scientific notation.

Lesson 2-1: Positive Integer Powers

Exponents represent the number of times that a number is multiplied by itself. The product $2\times2\times2\times2\times2$ involves multiplying 2 by itself 5 times. It can be written as either $2\times2\times2\times2\times2$ or as 2^5. They both mean the same thing, but using exponents takes less room and avoids the problem of miscounting the number of times that 2 appears in the product. In the expression 2^5, the number 5 is called the **exponent**, or the **power**, and the number 2 is called the **base**. In the expression 5^9, the exponent is 9 and the base is 5.

2

EXPONENTS

When dealing with exponential expressions, it is important to correctly identify the base and the exponent. If everything is positive, this task is straightforward. For example, in the expression 3^8, the base is 3 and the exponent is 8. This expression is shorthand for the product of 3 times itself 8 times. We can expand 3^8 and write $3^8 = 3 \times 3 \times 3 \times 3 \times 3 \times 3 \times 3 \times 3$. Exponents are a cool invention because they enable us to write numbers that are even larger than we can comprehend.

Identifying the base of an exponent becomes more difficult when negative numbers are involved. When combining negative numbers and exponents, parentheses become very important. For example, if the base is –4 and the exponent is 6, you would write this number as $(-4)^6$. The parentheses make it absolutely clear that the base is negative, for if you leave off the parentheses and write -4^6 what you are actually writing is the number $-(4^6)$. In other words, the base, 4, is multiplied by itself 6 times and then made into a negative number. So -4^6 is the negative of 4^6, and 4^6 is the number with base 4 and exponent 6. There is a big difference between $(-4)^6$ and -4^6. It's worth expanding both expressions just to make the point:

$$(-4)^6 = (-4)(-4)(-4)(-4)(-4)(-4) = 4{,}096$$
$$-4^6 = -4 \times 4 \times 4 \times 4 \times 4 \times 4 = -4{,}096$$

Notice that the absolute value of these two numbers is the same; it is their signs that are different.

We can determine whether an exponential expression will be positive or negative by observing patterns. Let's expand powers of –1:

$$(-1)^1 = -1$$
$$(-1)^2 = (-1)(-1) = 1$$
$$(-1)^3 = [(-1)(-1)](-1) = 1 \times (-1) = -1$$
$$(-1)^4 = [(-1)(-1)][(-1)(-1)] = 1 \times 1 = 1$$
$$(-1)^5 = [(-1)(-1)][(-1)(-1)](-1) = 1 \times 1 \times (-1) = -1$$

Notice that when the exponent is even, the answer is 1 and when the exponent is odd the answer is –1. The expression $(-4)^6$ involves a negative base and an even exponent, and the result was a positive number. Whenever you are multiplying a long string of negative and positive numbers, the easiest way to determine the overall sign is to count the number of negatives involved in the product.

- ◘ If the product involves an *even* number of negatives, the overall sign of the product will be positive.

- ◘ If the product involves an *odd* number of negatives, the overall sign of the product will be negative.

Example 1

Will the following products be positive or negative?

a. $(-6)^8$

b. $(-3)^3$

c. $-3 \times (-6)^8$

d. -2×3^3

e. $3 \times (-2)^4$

f. $-5 \times (-3)^3$

Solution:

a. The overall product will be positive: there are an even number (8) of negative signs in the product.

b. The overall product will be negative: there are an odd number (3) of negative signs in the product.

c. The overall product will be negative: there are 9 negative signs in the product (one from the –3 and 8 from $(-6)^8$.

d. The overall product will be negative: there is only one negative sign in the product.

2

EXPONENTS

e.   The overall product will be positive: there are 4 negative signs involved in the product (from $(-2)^4$).

f.   The overall product will be positive: there are 4 negative signs involved in the product (one from the –5 and 3 from $(-3)^3$).

With the introduction of exponents we will need to revisit our order of operations. Exponential expressions involve repetitive multiplication, and multiplication and division are done right after any instructions in parentheses. It should not surprise you to learn that exponentiation scores high on the order of operations. Parentheses still come first, though. Our expanded order of operations is now as follows:

▣ Parentheses

▣ Exponentiation

▣ Multiplication and division read left to right

▣ Addition and subtraction read left to right

Example 2

Use the order of operations to evaluate the following expressions:

a. 3×5^2

b. -2×3^3

c. $3 \times (-2)^4$

d. $-5 \times (-3)^3$

Solution:

a. $3 \times 5^2 = 3 \times 25 = 75$

b. $-2 \times 3^3 = -2 \times 27 = -54$

c.   First, find $(-2)^4$: $(-2)^4 = (-2) \times (-2) \times (-2) \times (-2) = 16$

Then use it to evaluate $3 \times (-2)^4$: $3 \times (-2)^4 = 3 \times 16 = 48$

d. First, find $(-3)^3$: $(-3)^3 = (-3) \times (-3) \times (-3) = -27$

Then use it to evaluate $-5 \times (-3)^3$: $-5 \times (-3)^3 = -5 \times (-27) = 135$

In general, an exponential expression is written as a^n and spoken as "a to the nth power." It is sometimes read as just "a to the nth" or as "the nth power of a." For example, 2^5 is read "two to the fifth power" or "two to the fifth."

There are some special powers that have specific names. For example, a^2 is read "a squared" and a^3 is read "a cubed." So 3^2 is read "three squared" and 3^3 is read "three cubed." The reason that the powers 2 and 3 have special names stems from their geometrical interpretation: the area of a square with side length a is $a \times a$ or a^2 (or "a squared"), and the volume of a cube with side length a is $a \times a \times a$ or a^3 (or "a cubed").

In the exponential expression a^n, a is the base and n is the exponent. Because of the way exponents have been defined, this expression only makes sense if n is a positive integer. We will have to expand our horizons in the next lesson if we want n to be anything other than a positive integer.

Lesson 2-1 Practice

Use the order of operations to evaluate the following expressions:

1. $-4 \times (-3)^2$

2. $4 \times (-5)^2$

3. $-3 \times (-2)^5$

Lesson 2-2: Rules for Exponents

Many mathematical rules were established by working out examples and looking for patterns. Now that we can deal with positive integer exponents, we can make some observations. Let's look at what happens when we multiply two exponential expressions with the same

2

EXPONENTS

base. For example, we can find the product $2^3 \times 2^5$ by expanding both factors and then writing the result using an exponent:

$$2^3 \times 2^5 = (2 \times 2 \times 2) \times (2 \times 2 \times 2 \times 2 \times 2)$$
$$= 2 \times 2 \times 2 \times 2 \times 2 \times 2 \times 2 \times 2 = 2^8$$

Notice that the exponents of the two terms in each of the products are 3 and 5, respectively, and the exponent in the result is 8, which is 3+5. This observation is actually our first rule for exponents, called the **product rule**: when you multiply two exponential expressions that have the same base, you add the exponents. This can be stated mathematically as:

$$a^m \times a^n = a^{m+n}$$

I will work out a few examples to illustrate how this product rule is used.

Example 1

Find the following products. Leave your answers as exponential expressions:

a. $3^5 \times 3^7$

b. $(-4)^6 \times (-4)^9$

c. $5^4 \times 5^3$

Solution:

a. $3^5 \times 3^7 = 3^{5+7} = 3^{12}$

b. $(-4)^6 \times (-4)^9 = (-4)^{6+9} = (-4)^{15}$

c. $5^4 \times 5^3 = 5^{4+3} = 5^7$

Next, let's explore what happens when you divide two exponential expressions with the same base. For example, we can evaluate the quotient $\dfrac{4^7}{4^4}$:

$$\frac{4^7}{4^4} = \frac{4 \times 4 \times 4 \times \cancel{4} \times \cancel{4} \times \cancel{4} \times \cancel{4}}{\cancel{4} \times \cancel{4} \times \cancel{4} \times \cancel{4}} = 4 \times 4 \times 4 = 4^3$$

Notice that every factor of 4 in the denominator cancels with a 4 in the numerator. The resulting exponent is just what you'd get if you subtracted the exponent in the denominator from the exponent in the numerator. This gives us our second rule for exponents, called the **quotient rule**: when you divide two exponential expressions with the same base you subtract the exponents. This can be stated mathematically as:

$$\frac{a^m}{a^n} = a^{m-n}$$

Here are some examples that will illustrate how the quotient rule can be used.

Example 2

Find the following quotients. Leave your answers as exponential expressions:

a. $\dfrac{3^7}{3^5}$

b. $\dfrac{(-4)^9}{(-4)^6}$

c. $\dfrac{5^4}{5^3}$

Solution:

a. $\dfrac{3^7}{3^5} = 3^{7-5} = 3^2$

b. $\dfrac{(-4)^9}{(-4)^6} = (-4)^{9-6} = (-4)^3$

c. $\dfrac{5^4}{5^3} = 5^{4-3} = 5^1 = 5$

So far, we have multiplied and divided exponential expressions. Now it's time to explore what happens when the base of an exponential expression is *itself* an exponential expression. In particular, let's examine the expression $(5^3)^4$. Using our order of operations, we must work out what is in parentheses first and then deal with the exponent:

$$(5^3)^4 = (5\times5\times5)^4 = (5\times5\times5)(5\times5\times5)(5\times5\times5)(5\times5\times5) = 5^{12}.$$

Notice that the final exponent is 12, because 5 is multiplied by itself 12 times. Amazingly enough, 12 is the product of 3 and 4, which are the exponents of each factor in the product. This generalizes into yet another rule for exponents, called the **power rule**: when you raise an exponential expression to a power, you multiply the exponents, or when you raise a power to a power you multiply the powers. We can write this mathematically as:

$$\left(a^m\right)^n = a^{m\times n}$$

I will work out a few examples to illustrate how the power rule is used. Keep in mind that in the future we will combine all three rules for simplifying exponents into a single problem.

Example 3

Evaluate the following. Leave your answers as exponential expressions:

a. $\left(3^4\right)^5$

b. $\left(\left(-4\right)^9\right)^4$

c. $\left(5^4\right)^3$

Solution:

a. $\left(3^4\right)^5 = 3^{4\times5} = 3^{20}$

b. $\left(\left(-4\right)^9\right)^4 = \left(-4\right)^{9\times4} = \left(-4\right)^{36}$

c. $\left(5^4\right)^3 = 5^{4\times3} = 5^{12}$

Here is your chance to practice what we have discussed. Be sure to check your answers before moving on to the next lesson.

Lesson 2-2 Practice

Evaluate the following. Leave your answers as exponential expressions:

1. $3^4 \times 3^8$

3. $\dfrac{4^8}{4^5}$

5. $\left(6^4\right)^8$

2. $(-5)^5 (-5)^3$

4. $\dfrac{(-2)^7}{(-2)^2}$

6. $\left(4^5\right)^3$

Lesson 2-3: Negative Integer Powers

Remember that there were several ways to represent the reciprocal of a non-zero number a. We could write the reciprocal of a as either $\dfrac{1}{a}$ or a^{-1}. Using this information we can interpret exponents that are negative integers. Applying the product rule to the expression $\left(3^{-1}\right)^4$ gives the expression 3^{-4}. Now, $\left(3^{-1}\right)^4$ can be thought of as the reciprocal of 3 raised to the 4th power. The reciprocal of 3 is $\dfrac{1}{3}$, so $3^{-4} = \left(3^{-1}\right)^4 = \left(\dfrac{1}{3}\right)^4$. We can then use the definition of exponents to evaluate $\left(\dfrac{1}{3}\right)^4$:

$$3^{-4} = \left(\dfrac{1}{3}\right)^4 = \left(\dfrac{1}{3}\right)\left(\dfrac{1}{3}\right)\left(\dfrac{1}{3}\right)\left(\dfrac{1}{3}\right) = \dfrac{1}{81}$$

On the other hand, the expression $\left(3^4\right)^{-1}$ will also simplify to 3^{-4} using the power rule for exponents: $\left(a^m\right)^n = a^{m \times n}$. We can evaluate 3^{-4} by first finding 3^4 and then taking its reciprocal:

$$3^{-4} = \left(3^4\right)^{-1} = 81^{-1} = \dfrac{1}{81}$$

From this, we see that $3^{-4} = \left(3^4\right)^{-1} = \left(3^{-1}\right)^4$

In general, if n is a positive integer, then

$$a^{-n} = \left(a^n\right)^{-1} = \left(a^{-1}\right)^n = \left(\dfrac{1}{a}\right)^n$$

The role of the negative sign in an exponent is to let you know whether the base belongs in the numerator or the denominator. Another way to look at it is that the negative exponent just means to take the reciprocal of the base. Keep in mind that:

- ☐ The reciprocal of a is $\dfrac{1}{a}$.

- ☐ The reciprocal of $\dfrac{1}{a}$ is a.

I will work out some examples to help illustrate how these rules are used.

Example 1

Evaluate the following exponential expressions:

   a. 2^{-3}

   b. $(-3)^{-2}$

   c. $\left(\dfrac{1}{4}\right)^{4}$

   d. $\left(-\dfrac{1}{2}\right)^{3}$

Solution:

   a. $2^{-3}=\left(2^{3}\right)^{-1}=8^{-1}=\dfrac{1}{8}$

   b. $(-3)^{-2}=\left[(-3)^{2}\right]^{-1}=9^{-1}=\dfrac{1}{9}$

   c. $\left(\dfrac{1}{4}\right)^{-4}=\left[\left(\dfrac{1}{4}\right)^{-1}\right]^{4}=4^{4}=64$

   d. $\left(-\dfrac{1}{2}\right)^{-3}=\left[\left(-\dfrac{1}{2}\right)^{-1}\right]^{3}=(-2)^{3}=-8$

    Take some time to work out the practice problems before moving on to the next set of rules. The problems will become more complex as a combination of rules will be necessary to simplify the expression. If

EXPONENTS

2

you practice using the rules individually, you will have an easier time using several of them in one problem.

Lesson 2-3 Practice

Evaluate the following exponential expressions:

1. 5^{-3} 2. $\left(\dfrac{1}{2}\right)^{-2}$ 3. $(-3)^{-3}$

Lesson 2-4: Zero as an Exponent

In order to interpret an exponential expression that has 0 as the exponent, we must revisit division of exponential expressions. Consider the ratio $\dfrac{a^n}{a^n}$, where $a \neq 0$. Using one of the properties of multiplication discussed earlier (the fact that any non-zero number divided by itself is equal to 1), we find that $\dfrac{a^n}{a^n} = 1$. Using the rules for dividing exponential expressions (when you divide two exponential expressions with the same base you subtract the exponents), we have $\dfrac{a^n}{a^n} = a^{n-n} = a^0$. In mathematics, consistency is crucial. So there's not much of a choice for how to interpret an exponential expression that has 0 as the exponent:

$$a^0 = 1$$

Using this idea, we can now explore two more exponential expressions: the expressions $\dfrac{1}{a^{-n}}$ and $\dfrac{1}{a^{n-m}}$. I will examine each one separately:

$$\frac{1}{a^{-n}} = \frac{a^0}{a^{-n}} = a^{0-(-n)} = a^n$$

$$\frac{1}{a^{n-m}} = \frac{a^0}{a^{n-m}} = a^{0-(n-m)} = a^{m-n}$$

The main idea in both expressions is that when you move an exponential expression from the numerator to the denominator, or vice versa, the net effect is that the sign of the exponent is changed.

Example 1

Evaluate exponential expression: $\dfrac{1}{3^{-2}}$

Solution: Use the first equation to simplify the expression:

$$\frac{1}{3^{-2}} = \frac{3^0}{3^{-2}} = 3^{0-(-2)} = 3^2 = 9$$

Example 2

Evaluate the exponential expression: $\dfrac{2^3}{2^{-2}}$

Solution: Use the second equation to simplify the expression:

$$\frac{2^3}{2^{-2}} = 2^{3-(-2)} = 2^5 = 32$$

Lesson 2-5: Powers of Quotients and Products

Now that we have discussed multiplying and dividing exponential expressions, we are ready to examine expressions that involve multiplication and exponentiation, like $(3a)^3$. Whenever you need to simplify exponential expressions it's always best to start with what the exponents mean and then make use of the properties of multiplication. Expand these expressions several times until you are comfortable with the interpretation of the short-hand notation. For example, the expression $(3a)^3$ means that $3a$ is multiplied by itself 3 times. We can expand the expression $(3a)^3$ and use the fact that multiplication is commutative and associative to rearrange the terms involved in the product:

$$(3a)^3 = (3a)(3a)(3a) = 3 \times 3 \times 3 \times a \times a \times a = 3^3 \times a^3$$

Notice that when you raise a product (in this case the product of 3 and a) to a power, in effect what you need to do is raise each term involved in the product to that power. It doesn't matter how many terms are involved in the product, or even if the terms in the products are themselves exponential expressions. Each term gets raised to that power. This rule is written as:

$$(a \times b)^n = a^n \times b^n$$

Example 1

Expand the following products:

a. $(4a)^2$

b. $(2b^2)^4$

c. $(-3a)^5$

Solution:

a. $(4a)^2 = 4^2 \times a^2 = 16a^2$

b. $(2b^2)^4 = 2^4 \times (b^2)^4 = 16b^{2\times4} = 16b^8$

c. $(-3a)^5 = (-1)^5 (3)^5 (a^5) = -243a^5$

Dealing with raising quotients to a power isn't any different. When you raise a quotient to a power, raise the numerator and the denominator to that power. It doesn't matter how complicated the numerator and denominator are. Use the appropriate rules to keep simplifying until there's nothing more you can do. If you take it one step at a time, you should be just fine. The rule for quotients is written as:

$$\left(\frac{a}{b}\right)^n = \frac{a^n}{b^n}$$

For example, the expression $\left(\frac{4}{a}\right)^3$ can be expanded and simplified as follows:

$$\left(\frac{4}{a}\right)^3 = \left(\frac{4}{a}\right)\left(\frac{4}{a}\right)\left(\frac{4}{a}\right) = \left(\frac{4^3}{a^3}\right)$$

Example 2

Evaluate the following:

a. $\left(\frac{a}{3}\right)^2$

b. $\left(\dfrac{2b^3}{3}\right)^5$

c. $\left(-\dfrac{4}{5a}\right)^3$

Solution:

a. $\left(\dfrac{a}{3}\right)^2 = \dfrac{a^2}{3^2} = \dfrac{a^2}{9}$

b. $\left(\dfrac{2b^3}{3}\right)^5 = \dfrac{\left(2b^3\right)^5}{3^5} = \dfrac{2^5 \times \left(b^3\right)^5}{243} = \dfrac{32b^{3\times5}}{243} = \dfrac{32b^{15}}{243}$

c. $\left(-\dfrac{4}{5a}\right)^3 = (-1)^3 \dfrac{4^3}{(5a)^3} = -\dfrac{64}{5^3 \times a^3} = -\dfrac{64}{125a^3}$

This is the last set of practice problems that focus on the rules for exponents. We will use all of these rules when I discuss monomials in Chapter 8. Practice applying this last rule, and be sure to check your answers before going on to the next lesson.

Lesson 2-5 Practice

Evaluate the following:

1. $(4b)^2$

2. $(-3c)^3$

3. $\left(2a^{-3}\right)^4$

4. $\left(\dfrac{b}{4}\right)^3$

5. $\left(\dfrac{-3}{a}\right)^2$

6. $\left(\dfrac{-4}{3a}\right)^3$

Lesson 2-6: Scientific Notation

Exponents are used to help us write down numbers whose magnitudes are either very large or very small. Archimedes, a famous Greek mathematician, managed to give a fairly accurate estimate of the number of grains of sand in the *universe* using the unwieldy number system available at the time (over 2000 years ago), but most people had trouble understanding his method. The distance between two atoms in a molecule is a very small number, and would have been challenging even for Archimedes to describe using his archaic number system. Since then, mathematicians have developed a number system, called scientific notation, that enables scientists to describe these extreme numbers very easily. Exponents are the basis for scientific notation.

Scientific notation is a standard way to write numbers using exponents with base 10. A number is written in scientific notation if it is of the form $a \times 10^n$ where a is a real number satisfying the inequality $1 \le |a| < 10$ and n is an integer.

The number 31,415 can be written using scientific notation as 3.1415×10^4. To convert a whole number like 31,415 to scientific notation, remember that an implied decimal point is at the end of the number. Move the decimal point to the *left* until there is only one digit to the left of it: 3.1415. Count how many places the decimal moved to the left. That number becomes the exponent or the power of 10. Because I moved the decimal point over 4 places to the left, the power of 10 is 4, and 31,415 is equivalent to 3.1415×10^4. It is easy to check the equivalence of 31,415 and 3.1415×10^4 by multiplying 3.1415 by 10^4. Recall that multiplying by 10 is equivalent to moving the decimal point one place to the right. Multiplying by 100 (or 10^2) is equivalent to moving the decimal point two places to the right. Multiplying by 10^n is equivalent to moving the decimal point n places to the right. So multiplying 3.1415 by 10^4 is equivalent to moving the decimal point 4 places to the right, resulting in the number 31,415.

We can also write numbers that are less than 1 in scientific notation. In this case, we will be moving the decimal point to the *right* until there is just one non-zero digit to the *left* of the decimal point. Each time that the decimal point is moved to the right the power of 10 decreases by 1. The number 0.005958 has a scientific notation representation

2

EXPONENTS

of 5.958×10^{-3}; the power of 10 is –3 because the decimal point was moved to the right 3 places: $0.005.958$. It doesn't matter if the number you are working with is positive or negative. The only thing that matters is the direction that you move the decimal point.

Example 1

Write the following numbers using scientific notation:

a. 386,450

b. –1,922.5

c. 0.00000646

d. –0.0004987

Solution:

a. $386,450 = 3.8645 \times 10^{5}$

b. $-1,922.5 = -1.9225 \times 10^{3}$

c. $0.00000646 = 6.46 \times 10^{-6}$

d. $-0.0004987 = -4.987 \times 10^{-4}$

One nice thing about working with numbers in scientific notation is that extremely large or extremely small positive numbers become more manageable. Specifically, multiplying and dividing these extreme numbers becomes *much* easier. When multiplying two numbers written in scientific notation, we will take advantage of the associative property of multiplication. Multiply the decimal numbers together and the exponential expressions together. Then make sure that your final answer is written in scientific notation. The same process holds when dividing two numbers written in scientific notation. The following examples should help illustrate the process.

EXPONENTS

2

Example 2

Evaluate the expression and write the result in scientific notation:

a. $(1.4 \times 10^4)(7.6 \times 10^3)$

b. $\dfrac{(1.6 \times 10^{-3})}{(3.2 \times 10^4)}$

c. $(2.0 \times 10^4)^3$

Solution:

a. $(1.4 \times 10^4)(7.6 \times 10^3) = (1.4 \times 7.6) \times (10^4 \times 10^3) = 10.64 \times 10^7$

$$= 1.064 \times 10^8$$

b. $\dfrac{(1.6 \times 10^{-3})}{(3.2 \times 10^4)} = \dfrac{1.6}{3.2} \times \dfrac{10^{-3}}{10^4} = 0.5 \times 10^{-7} = 5 \times 10^{-8}$

c. $(2 \times 10^4)^3 = (2^3)(10^4)^3 = 8 \times 10^{12}$

Lesson 2-6 Practice

1. Write the following numbers in scientific notation:

   a. 432,614

   b. 0.00482

2. Evaluate the following expressions and write your answer in scientific notation:

   a. $(3.42 \times 10^3)(2.87 \times 10^5)$

   b. $(3.2 \times 10^{-3})^2$

   c. $\dfrac{2.6 \times 10^4}{5.2 \times 10^{-2}}$

2

EXPONENTS

Answer Key
Lesson 2-1

1. -36 2. 100 3. 96

Lesson 2-2

1. 3^{12} 3. 4^3 5. 6^{32}

2. $(-5)^8$ 4. $(-2)^5$ 6. 4^{15}

Lesson 2-3

1. $\dfrac{1}{125}$ 2. 4 3. $-\dfrac{1}{27}$

Lesson 2-5

1. $16b^2$ 3. $16a^2$ or $\dfrac{16}{a^2}$ 5. $\dfrac{9}{a^2}$

2. $-27c^3$ 4. $\dfrac{b^3}{64}$ 6. $\dfrac{-64}{27a^3}$

Lesson 2-6

1.   a. 4.32614×10^5

     b. 4.82×10^{-3}

2.   a. 9.8154×10^8

     b. 1.024×10^{-5}

     c. 5×10^5

3

Equations and Equality

So far we have been working with algebraic *expressions*. You can think of expressions as being fragments of a sentence. In this chapter we will combine algebraic expressions to form equations, and we will practice solving these equations. Every equation has two sides, and there are rules in place that dictate what you can and cannot do to an equation.

Lesson 3-1: Equations

An **equation** is a statement that two expressions are equal. The two expressions that make up an equation are separated by the symbol "=" which is called an equal sign. Usually equations will contain variables, but they don't have to. Remember that a variable is a symbol that can be replaced by a variety of different numbers. Most of the time we use letters of the alphabet to represent variables, but technically you can use a smiley face or any other symbol to represent a variable.

You have already seen several examples of equations in this book. The statements $2^3 = 2 \cdot 2 \cdot 2$ and $a^m \cdot a^n = a^{m+n}$ are two examples of equations that you saw in Chapter 2. The first equation does not contain any variables, and the second equation contains 3 variables. Equations can have numbers and variables on both sides. The equations can be simple or they can be complicated. The important thing about an equation is that whatever expression appears on the left is exactly equal to whatever expression appears on the right. The two expressions don't have to look alike, but they must be equivalent.

Usually, when you are given an equation with only one variable your job will be to "solve the equation." This means that you will need to determine the specific numerical value (or values) of the variable that "satisfy the equality," or make the equation true. This process often involves moving all of the numbers to one side of the equation and all of the terms involving variables to the other side. When moving terms in an equation around, you must be sure to obey the rules.

An equation may contain terms that involve both a variable and a constant, like the term $3x$. The constant that is in front of the variable is called the **coefficient** of the variable. The expression $3x$ means the same thing as $3 \cdot x$ or $x + x + x$, and the expression $5x$ means the same thing as $5 \cdot x$ or $x + x + x + x + x$.

Lesson 3-2: Equality

I need to spend a few minutes talking about equality. Equality is an important symbol and concept used in mathematics. The concept of equality has very special properties. It is used to relate seemingly distinct objects, like the expressions 2^3 and $(10 - 8) \times 2 + 4$, and is called a relation. A **relation** is something that compares two objects. For example, if I had 5 one-dollar bills and you had 1 five-dollar bill then we would have an equal amount of money, even though the form of our money is different. If my son broke his piggy bank and discovered that he had 20 quarters, he would have the same amount of money as each of us, even though his money would weigh significantly more than ours did and would be better suited to be spent in an arcade. This example helps illustrate three important properties of equality.

- ◻ Equality is **reflexive**. This means that any object is equal to itself. This idea is almost self-evident. To illustrate the reflexive property mathematically we write $a = a$.
  For example, $3 = 3$.

- ◻ Equality is **symmetric**. This means that if object A is equal to object B, then object B is equal to object A. Basically it means that the order in which we equate objects doesn't matter. It doesn't matter whether object A appears on the left or the right hand side of the =. To state this idea mathematically we write: if $a = b$ then $b = a$.
  For example, if $2 \cdot 3 = 6$ then $6 = 2 \cdot 3$.

- ◻ Equality is **transitive**. This means that if object A is equal to object B, and object B is equal to object C, then object A is equal to object C. Object B is just an intermediate object that can be eliminated in the comparison; we can cut to the chase and set object A equal to object C. This can be expressed mathematically as: if $a = b$ and $b = c$, then $a = c$.
  For example, if $10 \cdot 2 = 20$ and $20 = 5 \cdot 4$, then $10 \cdot 2 = 5 \cdot 4$.

Because these three properties are so important in mathematics, we call any relation that has all three of these properties an **equivalence relation**. It is important to note that *every* equivalence relation (just like equality) has all three of these properties. There are other relations besides equality that we will use later on that are not equivalence relations. Those relations will be missing at least one of the three required properties: either the reflexive, the symmetric, or the transitive property.

Lesson 3-3: Algebraic Properties of Equality

As I mentioned earlier, there are rules to obey when working with equations. Keep in mind that the concept of equality is like a balanced scale. Whatever you have on the left side of the equal sign exactly equals whatever you have on the right. You are not allowed to tip the scales and favor one side over another side. Whatever you do to one side of the equation you must also do to the other side of the equation.

The first algebraic property of equality is known as the **addition property of equality**. It states that:

$$\text{If } a = b, \text{ then } a + c = b + c$$

We did not favor one side of the equality over the other side; the quantity c was added to both sides so that the balance is maintained. Looking back on the example with money, if you have a five-dollar bill and I have 5 one-dollar bills then we each have the same amount of money. If a kind stranger gives us each a twenty-dollar bill, then we are both still equally rich; we both have a total of $25. Notice that in the addition property of equality, c can be a positive or a negative number. If I have 5 one-dollar bills and my son has 20 quarters, and we both decide to spend $2 on an ice-cream cone, then we will both end up with $3. The addition property of equality is important because it allows us to add (or subtract) the same number from both sides of an equation without changing the validity of that statement.

The second algebraic property of equality is known as the **multiplication property of equality**. It states that:

$$\text{If } a = b, \text{ then } a \cdot c = b \cdot c$$

Again, one side is not favored over the other side; both sides are treated equally because both sides are being multiplied by c. Let's go back to our example with money. If you have a five-dollar bill and I have 5 one-dollar bills, and we both double our money in a bet, then we both now have $10. Notice again that with this property it doesn't matter if c is a positive or a negative number. I also didn't address whether c was greater than 1 or less than 1. That's because it doesn't matter. I can multiply (or divide) both sides of an equation by the same number and not change the validity of the statement.

These properties are certainly interesting, but, more importantly, they are crucial in your success in solving algebraic equations. We'll see how they are used in the next section.

Lesson 3-4: Solving Linear Equations in One Step

Linear equations are equations that involve only constants and one variable that is only raised to the first power. To solve a linear equation you need to find the specific value that the variable has to be in order to make the equation true. Finding the value of the variable requires you to isolate the variable on one side of the equal sign and have a number on the other side of the equal sign. In other words, if your linear equation involves the variable x your goal is to write an equation of the form $x = $ a number.

We can use the two algebraic properties of equality to help us solve linear equations. There are key transformations that you will use over and over to isolate the variable. When you use these transformations you will produce an equation that has the same solutions as the original equation. In other words, you will be transforming the original equation into a different, equivalent equation whose solution is easier to see. Remember that the goal is to get an equation of the form $x = a$, where a is some number; you can't get an easier equation than that to solve.

The first four transformations that we will practice are described in the table on page 64.

Anything that you do to one side of the equation must also be done on the other side of the equation. This ensures that balance is maintained. Remember that the goal is to determine the value of the variable that makes the equation true. In order to solve for the variable, you must get the variable by itself. These transformations help you isolate the variable.

3 EQUATIONS AND EQUALITY

Transformation	Original Equation	Transformation	New Equation
Interchange the sides of the equation	$7 = x$	Interchange	$x = 7$
Simplify one or both sides	$x = 6 + 9$	Simplify	$x = 15$
Add the same number to each side and simplify	$x - 5 = 9$	Add 5 to both sides and simplify	$x - 5 + 5 = 9 + 5$ $x = 14$
Subtract the same number from each side and simplify	$x + 7 = 12$	Subtract 7 from both sides and simplify	$x + 7 - 7 = 12 - 7$ $x = 5$

Example 1

Solve the equation: $x - 5 = 9$

Solution:

In order to isolate the variable, we need to get rid of the 5. To do this, add 5 to both sides and simplify:

$$x - 5 = 9$$

Add 5 to both sides: $\qquad\qquad x - 5 + 5 = 9 + 5$

Simplify: $\qquad\qquad\qquad\qquad x = 14$

The last step is to check our work: substitute $x = 14$ into our original equation and simplify. Is $14 - 5$ equal to 9? Yes. We can be confident that our answer is correct.

EQUATIONS AND EQUALITY

3

Example 2

Solve the equation: $x + 8 = -10$

Solution:

Subtract 8 from both sides and simplify:

$$x + 8 = -10$$

Subtract 8 from both sides: $\qquad x + 8 - 8 = -10 - 8$

Simplify: $\qquad x = -18$

Let's check our work. Substitute $x = -18$ into the original equation: Is $-18 + 8$ equal to -10? Yes.

Example 3

Solve the equation: $x - |-3| = 4$

Solution:

In this case we must first simplify the equation by evaluating the absolute value of -3. Then we need to isolate the variable by transforming our equation:

$$x - |-3| = 4$$

Evaluate $|-3|$: $\qquad x - 3 = 4$

Add 3 to both sides: $\qquad x - 3 + 3 = 4 + 3$

Simplify: $\qquad x = 7$

Checking our work we see that $7 - |-3| = 7 - 3$, and $7 - 3$ is equal to 4, so our answer is correct.

Transforming equations using addition and subtraction is just the beginning. We can also transform equations using multiplication and division. You'll want to transform equations using multiplication and division whenever the coefficient in front of the variable is any number other than 1. Remember, the goal is to get the variable by itself so that the equation states what the variable is equal to.

3

EQUATIONS AND EQUALITY

Transformation	Original Equation	Transformation	New Equation
Multiply each side of the equation by the same number	$\frac{1}{2}x = 5$	Multiply both sides of the equation by 2 and simplify	$2 \cdot \left(\frac{1}{2}x\right) = 2 \cdot 5$ $x = 10$
Divide each side of the equation by the same number	$3x = 12$	Divide both sides of the equation by 3 and simplify	$\frac{3x}{3} = \frac{12}{3}$ $x = 4$

You can also look at the division transformation in terms of multiplication. If you start with the equation $3x = 12$, you can multiply both sides of the equation by the reciprocal of the coefficient in front of the variable. In this case, the coefficient in front of the variable is 3, and its reciprocal is $\frac{1}{3}$. So, in order to solve the equation $3x = 12$ we could multiply both sides of this equation by $\frac{1}{3}$:

$$3x = 12$$

Multiply both sides by $\frac{1}{3}$: $\qquad\qquad \frac{1}{3} \cdot (3x) = \frac{1}{3} \cdot 12$

(or you could divide both sides of the equation by 3)

Simplify: $\qquad\qquad\qquad\qquad\qquad x = 4$

This perspective is helpful when the coefficient in front of the variable is not an integer.

Example 4

Solve the equation: $12 = -\frac{1}{4}x$

Solution:

You may have gotten used to seeing the variable on the left rather than on the right. Don't let that bother you; switch them around if it makes you feel better. Then multiply both sides of the equation by –4 and simplify:

$$12 = -\frac{1}{4}x$$

Interchange the sides of the equation:

$$-\frac{1}{4}x = 12$$

Multiply both sides of the equation by –4:

$$(-4)\cdot\left(-\frac{1}{4}x\right) = (-4)\cdot 12$$

Simplify:

$$x = -48$$

The last step is to check our work. Is $\left(-\frac{1}{4}\right)\cdot(-48)$ equal to 12? Yes it is.

Example 5

Solve the equation: $4x = -3$

Solution:

Divide both sides of the equation by 4 and simplify:

$$4x = -3$$

Divide both sides of the equation by 4:

$$\frac{4x}{4} = \frac{-3}{4}$$

Simplify:

$$x = \frac{-3}{4}$$

Of course you could have just multiplied both sides of the equation by $\frac{1}{4}$. The last step is to check our work. Is $4\cdot\left(-\frac{3}{4}\right)$ equal to –3? Yes.

Example 6

Solve the equation: $-\frac{3}{5}x = 6$

Solution:

Multiply both sides of the equation by the reciprocal of $-\frac{3}{5}$ (which is $-\frac{5}{3}$) and simplify:

$$-\frac{3}{5}x = 6$$

Multiply both sides of the equation by $-\frac{5}{3}$: $\left(-\frac{5}{3}\right)\cdot\left(-\frac{3}{5}x\right) = \left(-\frac{5}{3}\right)\cdot 6$

Simplify: $x = -10$

Finally, check our work. Is $\left(-\frac{3}{5}\right)\cdot(-10)$ equal to 6? Yes.

The best way to determine whether or not you have the hang of the subject matter is to take things into your own hands. Work out the following practice problems to see if you understand the concepts we just discussed. We will be building on these problem-solving techniques in the next lesson, so it is important that you work out some problems on your own before moving on. Be sure to check your answers at the end of this chapter.

Lesson 3-4 Practice

Solve the following equations:

1. $x + 5 = 9$
2. $x - 3 = 2$
3. $x + |-8| = 10$

4. $5 = -\frac{1}{3}x$
5. $5x = -3$
6. $-\frac{4}{5}x = 2$

Lesson 3-5: Solving Linear Equations Using Several Steps

Solving the problems in this section will involve applying the individual techniques discussed in the last lesson. The equations to solve will be more complicated, but if you take it one step at a time and remember to breathe, these problems shouldn't throw you for a loop.

In general, when you want to solve a linear equation involving one variable, the first thing you need to do is gather all of the terms that involve the variable over to one side of the equation and move all of the terms that don't involve the variable over to the other side of the equation. If, after combining all of the terms together, the coefficient in front of the variable is a number other than 1, you will need to multiply both sides of the equation by the reciprocal of the coefficient in front of the variable. Once you have done that, you should have an explicit equation for what numerical value the variable has to be. The last step is to check your work (using the original problem statement) to make sure that your answer is correct. This last step is the one that is most often skipped, but it's one that you really should get into the habit of doing. By checking your work you will know whether your answer is correct or not. If it's not, you'll have to go back and start over.

Example 1

Solve the equation: $\frac{1}{3}x + 4 = -3$

Solution:

To isolate the variable, you must move the 4 that appears on the left side of the equation over to the right side. This can be done by subtracting 4 from both sides of the equation and simplifying:

$$\frac{1}{3}x + 4 = -3$$

Subtract 4 from both sides:
$$\frac{1}{3}x + 4 - 4 = -3 - 4$$

Simplify:
$$\frac{1}{3}x = -7$$

Next, we want the coefficient in front of the variable to be 1. You can transform this equation by multiplying both sides of this equation by 3 and then simplifying:

$$\frac{1}{3}x = -7$$

Multiply both sides by 3:

$$3 \cdot \left(\frac{1}{3}x\right) = 3 \cdot (-7)$$

Simplify:

$$x = -21$$

Finally, check your answer. Is $\frac{1}{3} \cdot (-21) + 4 = -7 + 4$ equal to –3? Yes.

Example 2

Solve the equation: $4x = 2x - 8$

Solution:

In this case there are terms with variables appearing on both sides of the equation. Collect all of the terms with variables on one side of the equation and then solve the equation using the same techniques described earlier. In order to get all of the variables on one side of the equation, subtract $2x$ from both sides and simplify:

$$4x = 2x - 8$$

Subtract $2x$ from both sides:

$$4x - 2x = 2x - 8 - 2x$$

Simplify:

$$2x = -8$$

Now that our variables are on one side and the numbers are on the other side we can solve the equation. Divide both sides of the equation by 2 and simplify:

$$2x = -8$$

Divide both sides by 2:

$$\frac{2x}{2} = \frac{-8}{2}$$

Simplify:

$$x = -4$$

The last step is to check our work.

Is $4 \cdot (-4) = -16$ equal to $2 \cdot (-4) - 8 = -8 - 8 = -16$? Yes.

Example 3

Solve the equation: $5x + 4 = 2x + 10$

Solution:

Gather the variables on one side of the equation and the numbers on the other side. Then solve the equation:

$$5x + 4 = 2x + 10$$

Subtract 4 from both sides: $\qquad 5x + 4 - 4 = 2x + 10 - 4$

Simplify: $\qquad 5x = 2x + 6$

Subtract $2x$ from both sides: $\qquad 5x - 2x = 2x + 6 - 2x$

Simplify: $\qquad 3x = 6$

Multiply both sides by $\dfrac{1}{3}$: $\qquad \dfrac{1}{3} \cdot (3x) = \dfrac{1}{3} \cdot 6$

Simplify: $\qquad x = 2$

Finally, we'll check our answer. If $x = 2$ then $5x + 4 = 5 \cdot 2 + 4 = 14$

Example 4

Solve the equation: $3x + 4(x - 1) = 10$

Solution:

First distribute the 4, then solve for x:

$$3x + 4(x - 1) = 10$$

Distribute the 4: $\qquad 3x + 4x - 4 = 10$

Simplify: $\qquad 7x - 4 = 10$

Add 4 to both sides: $\qquad 7x - 4 + 4 = 10 + 4$

Simplify: $\qquad 7x = 14$

Divide both sides by 7: $\qquad \dfrac{7x}{7} = \dfrac{14}{7}$

Simplify: $\qquad x = 2$

Finally, check your answer.

If $x = 2$, then $3 \cdot 2 + 4(2 - 1) = 6 + 4 \cdot 1 = 10$. So our answer is correct.

Example 5

Solve the equation $4x - 2(x - 3) = 2$.

Solution:

First distribute the -2 carefully, then solve for x:

$$4x - 2(x - 3) = 2$$

Distribute the 2: $\qquad 4x - 2x + 6 = 2$

Simplify: $\qquad 2x + 6 = 2$

Subtract 6 from both sides: $\qquad 2x + 6 - 6 = 2 - 6$

Simplify: $\qquad 2x = -4$

Divide both sides by 2: $\qquad \dfrac{2x}{2} = \dfrac{-4}{2}$

Simplify: $\qquad x = -2$

Finally, check your answer.

If $x = -2$, then $4(-2) - 2(-2 - 3) = -8 - 2(-5) = -8 + 10 = 2$.

Our answer is correct.

Example 6

Solve the equation: $10 = \dfrac{5}{3}(x + 2)$

Solution:

Rewrite the equation so that the variable is on the left. You can either distribute the $\dfrac{5}{3}$ or multiply both sides of the equation by its reciprocal and solve for x:

$$10 = \frac{5}{3}(x + 2)$$

Interchange the sides of the equation: $\qquad \dfrac{5}{3}(x + 2) = 10$

Multiply both sides by $\dfrac{3}{5}$: $\qquad \dfrac{3}{5} \cdot \dfrac{5}{3}(x + 2) = \dfrac{3}{5} \cdot 10$

Simplify: $\qquad (x + 2) = 6$

Subtract 2 from both sides: $\qquad\qquad x+2-2=6-2$

Simplify: $\qquad\qquad\qquad\qquad\qquad x=4$

All that remains is to check our answer.

If $x=4$ then $\frac{5}{3}(4+2)=\frac{5}{3}\cdot 6=10$, so our answer is correct.

There are potential problems that you have to look out for when you are working with an equation that has more than one term involving the variable. One of the potential problems is that a solution to the equation may not exist. For example, the equation $x=x+2$ has no solution, because there is no number that is equal to itself plus 2. If you subtract x from both sides of the equation and simplify, you get:

$$x=x+2$$
$$x-x=x+2-x$$
$$0=2$$

Since this last equation is absurd, so is our original equation. Absurd equations have no solution.

Another potential problem is that there may not be a unique solution to the equation. For example, the equation $2(x+1)=2x+2$ is true regardless of what value of x you use. If you distribute the 2 on the left side of the equation you get $2x+2=2x+2$; $2(x+1)$ is always equal to $2x+2$. You can try any value of x and see that it works. In this case there are infinitely many solutions to the equation.

Here is your chance to put all of the pieces together. Take your time, and follow the same systematic approach that I used in the examples. Check your answers before moving on to the next lesson.

Lesson 3-5 Practice

Solve the following equations:

1. $\frac{2}{3}x-3=8$

2. $10x=3x-14$

3. $7x+5=2x-10$

4. $4x+2(x-1)=7$

5. $7x-2(x-4)=-2$

6. $12=\frac{4}{5}(x-2)$

3
EQUATIONS AND
EQUALITY

Lesson 3-6: Equations Involving Absolute Value

The next wrench that will be thrown into the system involves absolute values. Remember that we have an equation for the absolute value of a number:

$$|a| = \begin{cases} -a & \text{if } a < 0 \\ a & \text{if } a \geq 0 \end{cases}$$

The way you remove the absolute value symbols in an expression depends on what is inside of the absolute value symbols. If the contents of the absolute value symbols is *positive*, you can just drop the symbols. If the contents of the absolute value symbols is *negative*, then you are allowed to drop the symbols only after you put a negative sign in front of whatever was inside. If there are *variables inside of the absolute value symbols*, you won't know whether the contents are positive or negative, so you must explore both possibilities. In doing so, you will generate two equations when you remove the absolute value symbols.

When working with an equation that involves a variable inside of an absolute value symbol, the first thing you must do is isolate the absolute value part of the equation. It doesn't matter whether there are numbers or variables outside of the absolute value symbols, everything must be moved to the other side. Once you have done this, use the definition of absolute value to write two equations, depending on whether the contents of the absolute value symbols are positive or negative. Generate two equations that you will need to solve (and check your answers).

Example 1

Solve the equation: $|x| = 3$

Solution:

Since the absolute value symbol is already isolated, we can write out our two equations and solve them:

$$x = 3 \qquad -x = 3$$
$$x = -3$$

Either $x = 3$ or $x = -3$, and we have two possibilities that we have to check. Is $|3|$ equal to 3? Yes. Is $|-3|$ equal to 3? Yes. So both choices work: either $x = 3$ or $x = -3$. This can also be written as $x = \pm 3$.

Example 2

Solve the equation: $|x + 5| = 3$

Solution:

Again, the absolute value symbol is already isolated. Generate the two equations and solve each one by subtracting 5 from both sides:

$$x + 5 = 3 \qquad\qquad\qquad -(x + 5) = 3$$
$$x + 5 - 5 = 3 - 5 \qquad\qquad x + 5 = -3$$
$$x = -2 \qquad\qquad\qquad x + 5 - 5 = -3 - 5$$
$$x = -8$$

So we have two solutions: $x = -2$ or $x = -8$. We must check each one. Is $|-2 + 5| = |3|$ equal to 3? Yes. Is $|-8 + 5| = |-3|$ equal to 3? Yes. So both answers are correct: $x = -2$ or $x = -8$.

Example 3

Solve the equation: $|x + 2| + 3 = 6$

Solution:

We first need to isolate the absolute value by subtracting 3 from both sides. Then we will generate two equations that we can solve:

$$|x + 2| + 3 = 6$$
$$|x + 2| = 3$$

$$x + 2 = 3 \qquad\qquad\qquad -(x + 2) = 3$$
$$x + 2 - 2 = 3 - 2 \qquad\qquad x + 2 = -3$$
$$x = 1 \qquad\qquad\qquad x + 2 - 2 = -3 - 2$$
$$x = -5$$

3

EQUATIONS AND EQUALITY

The two solutions are $x = 1$ and $x = -5$.

We need to check each solution. Is $|1 + 2| + 3 = |3| + 3$ equal to 6? Yes. Is $|-5 + 2| + 3 = |-3| + 3$ equal to 6? Yes.

So both solutions are correct: $x = 1$ or $x = -5$.

Example 4

Solve the equation: $2|x + 1| = 3$

Solution:

Again, we need to isolate the absolute value by dividing both sides of the equation by 2. Once we have done that we can generate the two equations and solve each one:

$$2|x + 1| = 3$$

$$\frac{2|x + 1|}{2} = \frac{3}{2}$$

$$|x + 1| = \frac{3}{2}$$

$$x + 1 = \frac{3}{2} \qquad\qquad -(x + 1) = \frac{3}{2}$$

$$x + 1 - 1 = \frac{3}{2} - 1 \qquad\qquad x + 1 = -\frac{3}{2}$$

$$x = \frac{1}{2} \qquad\qquad x + 1 - 1 = -\frac{3}{2} - 1$$

$$x = -\frac{5}{2}$$

Let's check each solution:

If $x = \frac{1}{2}$, we have $2\left|\frac{1}{2} + 1\right| = 2\left|\frac{3}{2}\right| = 2 \cdot \frac{3}{2} = 3$,

and if $x = -\frac{5}{2}$ we have $2\left|-\frac{5}{2} + 1\right| = 2\left|-\frac{3}{2}\right| = 2 \cdot \frac{3}{2} = 3$,

so both of our answers work: $x = \frac{1}{2}$ or $x = -\frac{5}{2}$.

Example 5

Solve the equation: $|2x+1|=5$

Solution:

The absolute value is already isolated, so we just have to generate our two equations, solve them and check our solutions:

$$2x+1=5 \qquad\qquad -(2x+1)=5$$
$$2x=4 \qquad\qquad\qquad 2x+1=-5$$
$$x=2 \qquad\qquad\qquad\qquad 2x=-6$$
$$\qquad\qquad\qquad\qquad\qquad x=-3$$

Now to check our solution.

If $x=2$ then $|2\cdot2+1|=|5|=5$ and if $x=-3$ then $|2\cdot(-3)+1|=|-5|=5$. Both solutions check out: $x=2$ or $x=-3$.

Example 6

Solve the equation: $2|x+2|+2=4$

Solution:

First isolate the absolute value:

$$2|x+2|+2=4$$
$$2|x+2|=2$$
$$|x+2|=1$$

Next, generate our two equations and solve them:

$$x+2=1 \qquad\qquad -(x+2)=1$$
$$x=-1 \qquad\qquad\qquad x+2=-1$$
$$\qquad\qquad\qquad\qquad x=-3$$

Finally, check our answers.

If $x=-1$ then $2|-1+2|+2=2|1|+2=4$ and

if $x=-3$ then $2|-3+2|+2=2|-1|+2=4$.

Both solutions work, so $x=-1$ or $x=-3$.

Here are a few practice problems to help solidify the techniques we just discussed. Be sure to check your answers in the appendix before you move on to the next chapter. We will revisit absolute values and build on these problems in Chapter 4.

Lesson 3-6 Practice

Solve the following equations:

1. $|x+8|=2$
2. $|x-1|+4=6$
3. $3|x-1|=5$
4. $|3x-1|=5$
5. $\dfrac{1}{2}|x+3|+1=3$

Answer Key
Lesson 3-4

1. $x = 4$ (subtract 5 from both sides)

2. $x = 5$ (add 3 to both sides)

3. $x = 2$

4. $x = -15$ (multiply both sides by –3)

5. $x = -\dfrac{3}{5}$ (divide both sides by 5)

6. $x = -\dfrac{5}{2}$ or $x = -2\dfrac{1}{2}$ (multiply both sides by $-\dfrac{5}{4}$)

Lesson 3-5

1. $x = \dfrac{33}{2}$ or $x = 16\dfrac{1}{2}$ (add 3 to both sides and then multiply by $\dfrac{3}{2}$)

2. $x = -2$ (subtract $3x$ from both sides and then divide by 7)

3. $x = -3$ (subtract $2x$ from both sides, subtract 5 from both sides, divide by 5)

4. $x = \dfrac{3}{2}$ or $x = 1\dfrac{1}{2}$ (distribute the 2 and collect terms)

5. $x = -2$

6. $x = 17$

   (either distribute the $\dfrac{4}{5}$ and collect terms, or multiply both sides by $\dfrac{5}{4}$)

Lesson 3-6

1. $x = -6$ or $x = -10$
   (solve the two equations $x + 8 = 2$ and $-(x + 8) = 2$)

2. $x = 3$ or $x = -1$
   (solve the two equations $x - 1 = 2$ and $-(x - 1) = 2$)

3. $x = \dfrac{8}{3}$ or $x = -\dfrac{2}{3}$

    (solve the two equations $x - 1 = \dfrac{5}{3}$ and $-(x-1) = \dfrac{5}{3}$)

4. $x = 2$ or $x = -\dfrac{4}{3}$

    (solve the two equations $3x - 1 = 5$ and $-(3x-1) = 5$)

5. $x = 1$ or $x = -7$

    (solve the two equations $x + 3 = 4$ and $-(x+3) = 4$)

4

Inequalities and Graphs

Up until now we have been working with equality. I spent a bit of time talking about equality as an equivalence relation, and then I dropped the subject. Recall that an equivalence relation is a relation that has three important properties. Those properties were the reflexive, symmetric, and transitive properties. It's time to introduce you to two new relations: less than and greater than. These two relations are opposites of each other. Whatever I say about one of these relations also holds for the other one.

An **inequality** is an algebraic statement that compares two algebraic expressions that may not be equal. I mentioned four basic inequalities in Chapter 1 and I'll summarize them on page 82.

A solution to an inequality is the collection of all numbers that produce a true statement when substituted in for the variable in the inequality.

We can examine which of these inequalities, if any, are equivalence relations. I will examine $<$ and $\leq$ in more detail.

Symbol	Meaning	Example
$<$	Is less than	$3 < 8$
$\leq$	Is less than or equal to	$3 \leq 8,\ 8 \leq 8$
$>$	Is greater than	$8 > 3$
$\geq$	Is greater than or equal to	$8 \geq 3,\ 8 \geq 8$

To determine whether $<$ (less than) is an equivalence relation, we need to check to see if it has all three required properties:

- **Reflexive:** Is $a < a$? In other words, is a number less than itself? No.
- **Symmetric:** If $a < b$ is $b < a$? No.
- **Transitive:** If $a < b$ and $b < c$, is $a < c$? Yes.

So $<$ is not an equivalence relation because it does not have the reflexive and symmetric properties.

To determine whether $\leq$ (less than or equal to) is an equivalence relation, we need to check to see whether it has all three required properties:

- **Reflexive:** Is $a \leq a$? In other words, is a number less than or equal to itself? Yes.
- **Symmetric:** If $a \leq b$ is $b \leq a$? Not always, so no.
- **Transitive:** If $a \leq b$ and $b \leq c$, is $a \leq c$? Yes.

So $\leq$ is not an equivalence relation because it does not have the symmetric property.

You may be wondering why I care so much about equivalence relations. Well, in order to solve inequalities we will need to establish some rules about how we are allowed to transform inequalities. There are some transformations that we are allowed to do with inequalities that we are not allowed to do with equalities. The difference in how we treat an equality versus an inequality stems from the fact that equality is an equivalence relation whereas inequalities are not.

Lesson 4-1: Properties of Inequalities

Remember that we discussed two algebraic properties of equality: the addition property of equality and the multiplication property of equality. Inequalities actually have *two* addition properties and a restricted multiplication property.

The first addition property of inequality states that:

$$\text{If } a > b, \text{ then } a + c > b + c$$

This property isn't really surprising. Let's go back to an example with money. Suppose that I have $20 and you have $10. If someone gives us each $5, then I will still have more money than you. I'll have $25 and you'll have $15. We will both be richer, but I'll still be richer than you.

The second addition property of inequality states that:

$$\text{If } a > b \text{ and } c > d, \text{ then } a + c > b + d$$

Again, this is not surprising. If I have $20 and you have $10, and if some kind person gives me $20 and you $10, then I will definitely still have more money than you.

Inequalities are more flexible than equalities, but there is a limit to what you can get away with. One thing that you are not allowed to do is disturb the inequality. In other words, either you give the same amount to both sides of the inequality or you give more to "haves" then you do to the "have nots." You cannot try to make up for any injustice by giving more to the "have nots." It may not seem fair, but that's why we call them inequalities.

The restricted multiplication property is also fairly clear. It states that

$$\text{If } a > b \text{ and } c > 0, \text{ then } a \cdot c > b \cdot c$$

The restriction that the number that you multiply by must be positive is very important, and worth looking at in more detail. Consider the inequality $0 < 1$. Suppose that we wanted to multiply both sides of this inequality by a negative number; let's multiply both sides of this inequality by –1. On the left-hand side of the inequality you would have $0 \cdot (-1) = 0$ and on the right-hand side of the inequality you would have $1 \cdot (-1) = -1$. Now, how does 0 compare to –1? Well, $0 > -1$. Notice

INEQUALITIES AND GRAPHS

4

that we started with $0 < 1$ and when we multiplied both sides of the inequality by –1 we end up with the inequality $0 > -1$. This leads us to an important rule about multiplying an inequality by a negative number: when you multiply an inequality by a negative number you must also remember to flip the inequality from < to > or vice versa. We can write this mathematically as:

$$\text{If } a < b \text{ and } c < 0, \text{ then } a \cdot c > b \cdot c$$

This flipping rule also holds when you divide both sides of an inequality by a negative number.

Lesson 4-2: Solving Inequalities in One Step

Solving an inequality is a lot like solving an equality. The goal is still to isolate the variable on one side of the inequality. The transformation rules for inequalities are similar to those for equality. The main difference in the transformations occur when you multiply or divide both sides of an inequality by a negative number. You must be sure to flip the inequality when you multiply (or divide) both sides of an inequality by a negative number.

The addition transformations are summarized in the table on the previous page. Notice that you no longer have the ability to interchange the sides of the inequality. That's because inequalities do not have the symmetric property.

Transformation	Original Equation	Transformation	New Equation
Simplify one or both sides	$x > 6 + 9$	Simplify	$x > 15$
Add the same number to each side and simplify	$x - 5 > 9$	Add 5 to both sides and simplify	$x - 5 + 5 > 9 + 5$ $x > 14$
Subtract the same number from each side and simplify	$x + 7 > 12$	Subtract 7 from both sides and simplify	$x + 7 - 7 > 12 - 7$ $x > 5$

Example 1

Solve the inequality: $x + 3 < 4$

Solution: Subtract 3 from both sides and simplify:

$x + 3 < 4$

$x + 3 - 3 < 4 - 3$

$x < 1$

The solution is all real numbers less than 1.

Example 2

Solve the inequality: $-3 > x - 6$

Solution: Add 6 to both sides and simplify:

$-3 > x - 6$

$-3 + 6 > x - 6 + 6$

$3 > x$

The solution is all real numbers less than 3.

The transformations involving multiplication and division are summarized in the table on page 86.

Example 3

Solve the inequality: $\dfrac{6}{11}x > 2$

Solution: Multiply both sides by $\dfrac{11}{6}$ and simplify:

$\dfrac{6}{11}x > 2$

$\dfrac{11}{6} \cdot \dfrac{6}{11}x > \dfrac{11}{6} \cdot 2$

$x > \dfrac{11}{3}$

The solution is all real numbers greater than $\dfrac{11}{3}$.

Transformation	Original Equation	Transformation	New Equation
Multiply both sides by the same *positive* number and simplify	$\frac{1}{2}x > 9$	Multiply both sides by 2 and simplify	$2 \cdot \left(\frac{1}{2}x\right) > 2 \cdot 9$ $x > 18$
Divide both sides by the same *positive* number and simplify	$2x > 8$	Divide both sides by 2 and simplify	$\frac{2x}{2} > \frac{8}{2}$ $x > 4$
Multiply both sides by the same *negative* number, reverse the inequality, and simplify	$-\frac{2}{3}x > 4$	Multiply both sides by $-\frac{3}{2}$, flip the inequality, and simplify	$\left(-\frac{3}{2}\right)\left(-\frac{2}{3}x\right) < \left(-\frac{3}{2}\right)4$ $x < -6$
Divide both sides by the same *negative* number, reverse the inequality, and simplify	$-3x > 4$	Divide both sides by –3, flip the inequality, and simplify	$\frac{-3x}{-3} < \frac{4}{-3}$ $x < -\frac{4}{3}$

Example 4

Solve the inequality: $4x > 5$

Solution: Divide both sides by 4 and simplify:

$4x > 5$

$\frac{4x}{4} > \frac{5}{4}$

$x > \frac{5}{4}$

The solution is all real numbers greater than $\frac{5}{4}$.

Example 5

Solve the inequality: $-3x < 9$

Solution: Divide both sides by –3, flip the inequality and simplify:

$$-3x < 9$$
$$\frac{-3x}{-3} > \frac{9}{-3}$$
$$x > -3$$

The solution is all real numbers greater than –3.

Example 6

Solve the inequality: $-\frac{1}{3}x > -2$

Solution: Multiply both sides by –3, flip the inequality and simplify:

$$-\frac{1}{3}x > -2$$
$$(-3)\left(-\frac{1}{3}x\right) < (-3)(-2)$$
$$x < 6$$

The solution is all real numbers less than 6.

These practice problems will give you an opportunity to get into the game. I have been having all of the fun solving problems, so now it is your turn. These problems illustrate the two steps used to solve inequalities. We will put them together and solve more complicated problems in the next lesson.

Lesson 4-2 Practice

Solve the following inequalities:

1. $x + 5 < 9$

2. $-2 > x - 10$

3. $\frac{4}{15}x > 6$

4. $3x > 8$

5. $-4x < 10$

6. $-\frac{1}{2}x > -5$

INEQUALITIES AND GRAPHS

4

Lesson 4-3: Solving Inequalities Using Several Steps

In general, when you want to solve an inequality involving one variable the first thing you need to do is move all of the terms that involve the variable to one side of the inequality and move all of the terms that don't involve the variable over to the other side. If, after combining all of the terms together the coefficient in front of the variable is a number other than 1, you will need to multiply both sides of the inequality by the reciprocal of the coefficient in front of the variable. If this sounds a lot like the technique we used to solve equalities using several steps in the last chapter, you would be right.

Example 1

Solve the inequality: $\frac{1}{3}x + 4 > -3$

Solution: Subtract 4 from both sides, then multiply by 3 and simplify.

$$\frac{1}{3}x + 4 > -3$$

Subtract 4 from both sides: $\frac{1}{3}x + 4 - 4 > -3 - 4$

Simplify: $\frac{1}{3}x > -7$

Multiply both sides by 3: $3 \cdot \left(\frac{1}{3}x\right) > 3 \cdot (-7)$

Simplify: $x > -21$

The solution is all real numbers greater than –21.

Example 2

Solve the inequality: $2x + 10 \geq 7(x + 1)$

Solution: First distribute the 7 on the right, then collect all the terms involving variables on one side and the numbers on the other. Finally solve for x:

$$2x+10 \geq 7(x+1)$$

Distribute the 7: $\qquad\qquad 2x+10 \geq 7x+7$

Subtract 10 from both sides: $\qquad 2x+10-10 \geq 7x+7-10$

Simplify: $\qquad\qquad\qquad -5x \geq -3$

Divide both sides by –5 and flip
the inequality: $\qquad\qquad \dfrac{-5x}{-5} \leq \dfrac{-3}{-5}$

Simplify: $\qquad\qquad\qquad x \leq \dfrac{3}{5}$

The solution is all real numbers less than or equal to $\dfrac{3}{5}$.

Example 3

Solve the inequality: $12 > -2x - 6$

Solution: Add 6 to both sides and then divide both sides by –2.
Remember to flip the inequality.

$$12 > -2x - 6$$

Add 6 to both sides: $\qquad\qquad 12+6 > -2x-6+6$

Simplify: $\qquad\qquad\qquad 18 > -2x$

Divide both sides by –2 and flip
the inequality: $\qquad\qquad \dfrac{18}{-2} < \dfrac{-2x}{-2}$

Simplify: $\qquad\qquad\qquad -9 < x$

The solution is all real numbers greater than –9.

Example 4

Solve the inequality: $10 \geq \dfrac{5}{3}(x+2)$

Solution: There are several ways to start this problem. You could
distribute the $\dfrac{5}{3}$ and then move the terms around. Or, you could
first multiply by $\dfrac{3}{5}$ and then simplify. I will work the problem out the
second way, and leave it to you to work it out the first way:

INEQUALITIES AND GRAPHS

4

$$10 \geq \frac{5}{3}(x+2)$$

Multiply both sides by $\frac{3}{5}$: $\qquad \frac{3}{5} \cdot 10 \geq \frac{3}{5} \cdot \frac{5}{3}(x+2)$

Simplify: $\qquad 6 \geq x+2$

Subtract 2 from both sides: $\qquad 6-2 \geq x+2-2$

Simplify: $\qquad 4 \geq x$

The solution is all real numbers less than or equal to 4.

Here is your chance to shine. Put all the pieces together and see what you can do. Be sure to check your answers before moving on to the next section.

Lesson 4-3 Practice

Solve the following inequalities:

1. $\frac{1}{5}x+1<2$

2. $3<-4x-2$

3. $2 \geq -\frac{1}{3}(x+2)$

4. $3x>5x-8$

5. $2x+4 \leq x-3$

6. $3-2(x+4) \geq 2x+11$

Lesson 4-4: Graphing Inequalities Using the Number Line

When we solved the inequalities in the last section, our answers were written algebraically. There wasn't a unique answer, but rather there were many real numbers that satisfied the inequality. We described the collection of these solutions with words, but a picture is worth a thousand words. It helps to understand the solution to an inequality by being able to visualize it. To do this, we will make use of the number line.

Zero is one of the most important points on a number line. The point 0 is called the origin. Negative numbers lie to the left of the origin, positive numbers lie to the right. To draw a number line, just draw a line with arrows pointing in both directions, and label a point 0. It's best to keep your number lines simple. Some people start labeling the integers around 0, but I recommend only labeling points that are pertinent. And at this point, the only important point is 0. Take a look at my number line shown in Figure 4.1.

0

Figure 4.1 The number line.

Any real number can serve as the starting point for a ray. A **ray** is half of a line. A ray on the number line looks like an arrow. A ray can either point to the left or to the right. It has a definite starting point, but it never ends. Remember the dots after the counting numbers indicated that the counting numbers never end. A ray is similar in that it goes on forever. A ray that points to the left will include all of the points to the left of, or less than, the starting point. A ray that points to the right will include all of the points to the right of, or greater than, the starting point. The direction that a ray points depends on the inequality used to describe the ray. The set of all points less than a given point is drawn as an arrow starting at the given point and pointing to the left. The set of all points greater than a given point is drawn as an arrow starting at the given point and pointing to the right.

The starting point of a ray may or may not be included. The relations < (less than) and > (greater than) are called **strict inequalities**. For strict inequalities, the starting point is *not* included in the ray. The relations ≤ (less than or equal to) and ≥ (greater than or equal to) are just referred to as inequalities, and for these inequalities the starting point *is* included in the ray (because of the possibility that both sides of the inequality are actually equal to each other). An **open ray** is a ray that does not include its starting point; a **closed ray** is a ray that does include its starting point.

To draw a ray, start with a number line and label your starting point. Once you have your starting point labeled, determine if your ray is open or closed. If it's open (meaning it involves the relations $<$ or $>$), draw an open circle at your labeled starting point. If it's closed (meaning it involves the relations $\leq$ or $\geq$), draw a filled-in circle at your labeled starting point. Then draw an arrow that points in the correct direction (to the left for $<$ or $\leq$ and to the right for $>$ or $\geq$) to represent your ray. Figure 4.2 shows the open ray representing the set of real numbers less than 2.

Figure 4.2: The open ray $x < 2$.

Figure 4.3 shows an open ray representing the set of real numbers *less than –2* and the closed ray representing the set of real numbers *less than or equal to –2*. Notice that the only difference between these two rays is the circle at –2: the open ray has an open circle and the closed ray has a closed (or filled in) circle. Both of these rays point to the left because we are describing points that are less than (or less than or equal to) –2.

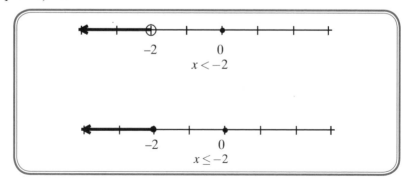

Figure 4.3: The open ray $x < -2$ and the closed ray $x \leq -2$.

Figure 4.4 shows an open ray representing the set of real numbers *greater than 4* and the closed ray representing the set of real numbers *greater than or equal to 4*. Notice that the only difference between these two rays is the circle at 4: the open ray has an open circle and the closed ray has a closed (or filled in) circle. Both of these rays point to the right because we are describing points that are greater than (or greater than or equal to) 4.

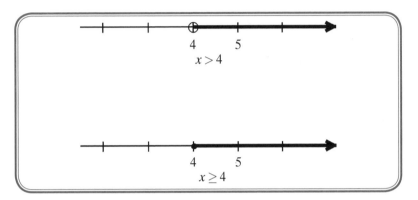

$$x > 4$$

$$x \geq 4$$

Figure 4.4: The open ray $x > 4$ and the closed ray $x \geq 4$.

In the last section we solved several inequalities and described the solution in words. We will take a minute to give a graphical solution to those same inequalities.

Example 1

Graph the solution to the inequality: $\dfrac{1}{3}x + 4 > -3$

Solution: The solution to this inequality is $x > -21$, which is the open ray shown in Figure 4.5.

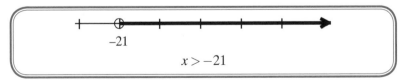

$$x > -21$$

Figure 4.5: The open ray $x > -21$.

Example 2

Graph the solution to the inequality: $2x + 10 \geq 7(x+1)$

Solution: The solution to this inequality is $x \leq \dfrac{3}{5}$, which is the closed ray shown in Figure 4.6.

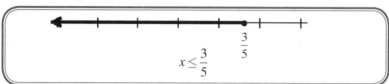

Figure 4.6: The closed ray $x \leq \dfrac{3}{5}$.

Example 3

Graph the solution to the inequality: $12 > -2x - 6$

Solution: The solution to this inequality is $-9 < x$, which is the open ray shown in Figure 4.7.

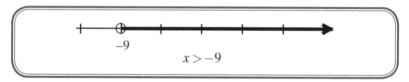

Figure 4.7: The open ray $x > -9$.

Example 4

Graph the solution to the inequality: $10 \geq \dfrac{5}{3}(x+2)$

Solution: The solution to this inequality is $4 \geq x$, which is the closed ray shown in Figure 4.8.

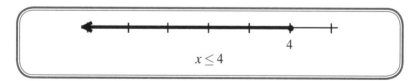

Figure 4.8: The closed ray $x \leq 4$.

Now it is your turn to practice solving these problems. If you get stuck, take a look back at the examples I worked. As you solve more problems you will start to recognize the similarities in how to approach them. The strategy I showed you can be applied to solve all of these problems.

Lesson 4-4 Practice

Graph the solutions to the following inequalities:

1. $\frac{1}{5}x + 1 < 2$

2. $3 < -4x - 2$

3. $2 \geq -\frac{1}{3}(x + 2)$

4. $3x > 5x - 8$

5. $2x + 4 \leq x - 3$

6. $3 - 2(x + 4) \geq 2x + 11$

Lesson 4-5: Graphing Compound Inequalities

Compound inequalities involve two inequalities. Since each individual inequality is graphed as a ray, graphing compound inequalities requires you to graph two rays. Depending on the inequalities involved, your solution will either consist of all of the points that lie on *either* ray or only those points that lie on *both* rays. In order to determine the points that lie on both rays you will need to focus on where they overlap.

Example 1

Graph the compound inequality: $x > 3$ or $x \leq 0$

Solution: Graph each inequality separately, as shown in Figure 4.9. Since you can either be greater than 3 or less than or equal to 0, the solution is the set of points that lie on either ray.

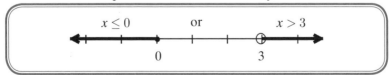

Figure 4.9. The set of all points $x > 3$ or $x \leq 0$.

Example 2

Graph the compound inequality: $x \leq 4$ and $x > 0$

Solution: Graph each inequality separately. Because your solution must include all points that satisfy both inequalities (as noted by the "and"), you are looking for where the two rays overlap (or intersect). The solution is shown in Figure 4.10.

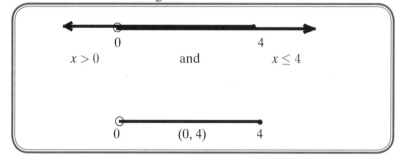

Figure 4.10: The set of all points $x \leq 4$ and $x > 0$.

In the event that you are looking at two rays that point in opposite directions and overlap, like you saw in Example 2, your solution is called an interval. An **interval** is part of a number line that has both a starting point and a stopping point. The starting point and the stopping point are called the **endpoints** of the interval. The endpoints of an interval may or may not be included. If both endpoints are included, the interval is called a **closed interval**. If neither endpoint is included,

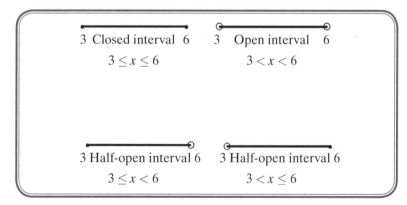

Figure 4.11: Closed, open, and half-open intervals.

the interval is called an **open interval**. If one endpoint is included and the other endpoint is not included, the interval is called either a **half-open interval** or a **half-closed interval**. Figure 4.11 on page 96 shows examples of the different types of intervals you will encounter here.

Lesson 4-5 Practice

Graph the following compound inequalities:

1. $x > 2$ and ≤ 5 2. $x < -2$ or $x > 1$ 3. $x \leq 4$ and $x \geq 0$

Lesson 4-6: Inequalities and Absolute Value

When you solved absolute value equations, you'll recall that you ended up creating two equations after removing the absolute value part of the equation. Well, it shouldn't surprise you that when you mix inequalities and absolute value symbols, you will end up creating two *inequalities* when you remove the absolute value symbols in the equation. Satisfying two inequalities in the same problem means that the solution will be a compound inequality. Hopefully you are beginning to see how we are building on what we have learned in earlier chapters and earlier sections. That's a large part of learning mathematics. We continually push to learn new skills, and then use those new skills to help develop and explore new ideas. Every skill you learn will form the foundation on which new knowledge is built. So let's solve some more problems.

Your approach to solving inequalities with absolute value should be almost identical to the approach you used to solve equations that involved absolute value. First, isolate the absolute value part of the inequality. Then create two inequalities and solve each one using the **rule for inequalities**:

$$|a| = \begin{cases} -a & \text{if } a < 0 \\ a & \text{if } a \geq 0 \end{cases}$$

Put the solutions to the two inequalities together to get the final solution.

Usually, when your equation involves an absolute value of something that is less than (or less than or equal to) a number, you will need to satisfy both of the inequalities that you generate and your answer will consist of the regions where the two rays overlap. If your equation involves an absolute value of something that is greater than (or greater than or equal to) a number, your solution will be all of the numbers that lie on one ray or the other.

The inequality $|ax+b|>c$ is used to create two inequalities, depending on whether $ax+b$ is positive or negative. If $ax+b$ is positive, then $|ax+b|=ax+b$ and the inequality $|ax+b|>c$ becomes $ax+b>c$. If $ax+b$ is negative, then $|ax+b|=-(ax+b)$ and the inequality $|ax+b|>c$ becomes $-(ax+b)>c$. The two inequalities $ax+b>c$ and $-(ax+b)>c$ represent the *initial meaning* of the inequality $|ax+b|>c$. If we multiply the inequality $-(ax+b)>c$ by -1, we get $ax+b<-c$. The two inequalities $ax+b>c$ or $ax+b<-c$ represent the *simplified meaning* of the inequality $|ax+b|>c$.

The inequalities $ax+b>c$ and $ax+b<-c$ represent two rays that do not overlap. The solution to the inequality $|ax+b|>c$ will be the set of points that satisfy one inequality or the other.

Inequality	Initial Meaning	Simplified Meaning
$|ax+b|>c$	$ax+b>c$ or $-(ax+b)>c$	$ax+b>c$ or $ax+b<-c$
$|ax+b|\geq c$	$ax+b\geq c$ or $-(ax+b)\geq c$	$ax+b\geq c$ or $ax+b\leq -c$
$|ax+b|<c$	$ax+b<c$ and $-(ax+b)<c$	$ax+b<c$ and $ax+b>-c$
$|ax+b|\leq c$	$ax+b\leq c$ and $-(ax+b)\leq c$	$ax+b\leq c$ and $ax+b\geq -c$

This same analysis is done for the other three types of inequalities you will encounter in this book, and the results are shown in the table on page 98. When I work out problems I will remove the absolute values using the initial meaning. Then I will carefully transform the inequality, flipping the inequality when necessary. I will include a graphic solution to each problem, to keep my graphing skills fresh.

Example 1

Solve the inequality $|x+2|<4$ and graph the solution.

Solution: The absolute value is already isolated. So all we need to do is generate our two inequalities by substituting for the absolute value symbols. We then solve both inequalities:

$$x+2<4 \qquad -(x+2)<4$$
$$x<2 \qquad x+2>-4$$
$$x>-6$$

The solution is the set of real numbers that are less than 2 and greater than –6. Figure 4.12 shows the solution graphically.

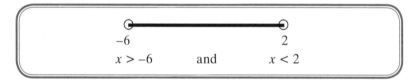

Figure 4.12: The graphical solution of $|x+2|<4$.

Example 2

Solve the inequality $|2x-1|+2\geq7$ and graph the solution.

Solution: First isolate the absolute value by subtracting 2 from both sides:

$$|2x-1|+2\geq7$$
$$|2x-1|\geq5$$

4

INEQUALITIES AND
GRAPHS

Next, generate the two inequalities and solve them both:

$$2x - 1 \geq 5 \qquad\qquad -(2x - 1) \geq 5$$
$$2x \geq 6 \qquad\qquad\qquad 2x - 1 \leq -5$$
$$x \geq 3 \qquad\qquad\qquad\quad 2x \leq -4$$
$$\qquad\qquad\qquad\qquad\qquad x \leq -2$$

The solution is the set of real numbers that are either greater than or equal to 3, or less than or equal to –2. Figure 4.13 shows the solution graphically.

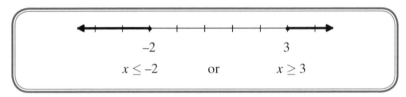

$$x \leq -2 \qquad \text{or} \qquad x \geq 3$$

Figure 4.13: The graphical solution of $|2x - 1| + 2 \geq 7$.

Lesson 4-6 Practice

Graph the solutions to the following inequalities.

1. $|x - 3| \geq 3$

2. $|3x + 2| \leq 8$

3. $|2x + 3| - 5 < 7$

4. $2|3x - 9| > 10$

Answer Key
Lesson 4-2

1. $x < 4$ (subtract 5 from both sides)

2. $x < 8$ (add 10 to both sides)

3. $x > \dfrac{45}{2}$ or $x > 22\dfrac{1}{2}$ (multiply both sides by $\dfrac{15}{4}$)

4. $x > \dfrac{8}{3}$ or $x > 2\dfrac{2}{3}$ (divide both sides by 3)

5. $x > -\dfrac{5}{2}$ or $x > -2\dfrac{1}{2}$ (divide both sides by −4 and flip the inequality)

6. $x < 10$ (multiply both sides by −2 and flip the inequality)

Lesson 4-3

1. $x < 5$ (subtract 1 from both sides and then multiply both sides by 5)

2. $x < -\dfrac{5}{4}$ or $x < -1\dfrac{1}{4}$ (add 2 to both sides, divide by −4 and flip the inequality)

3. $x \geq -8$ (multiply both sides by −3 and flip the inequality, then subtract 2)

4. $x < 4$

5. $x \leq -7$

6. $x \leq -4$

Lesson 4-4

1. $x < 5$

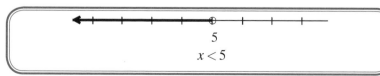

$x < 5$

2. $x < -\dfrac{5}{4}$ or $x < -1\dfrac{1}{4}$

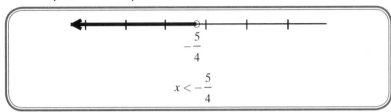

$-\dfrac{5}{4}$

$x < -\dfrac{5}{4}$

3. $x \geq -8$

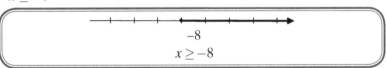

-8

$x \geq -8$

4. $x < 4$

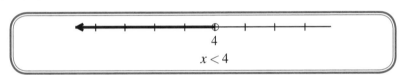

4

$x < 4$

5. $x \leq -7$

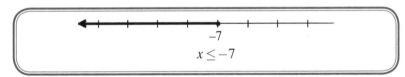

-7

$x \leq -7$

6. $x \leq -4$

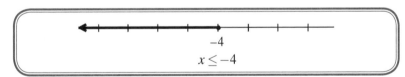

-4

$x \leq -4$

Lesson 4-5

1. $x > 2$ and $x \leq 5$

2 $\qquad$ 5

$x > 2$ $\qquad$ and $\qquad$ $x \leq 5$

2. $x < -2$ or $x > 1$

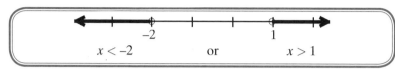

$x < -2$ \qquad or \qquad $x > 1$

3. $x \leq 4$ and $x \geq 0$

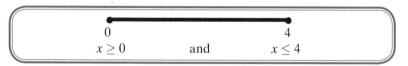

$x \geq 0$ \qquad and \qquad $x \leq 4$

Lesson 4-6

1. $x \geq 6$ or $x \leq 0$

    (solve the two inequalities $x - 3 \geq 3$ and $-(x-3) \geq 3$)

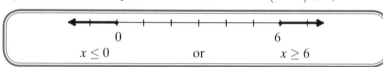

$x \leq 0$ \qquad or \qquad $x \geq 6$

2. $x \geq -\dfrac{10}{3}$ and $x \leq 2$

    (solve the two inequalities $3x + 2 \leq 8$ and $-(3x+2) \leq 8$)

$-\dfrac{10}{3}$ \qquad\qquad 2

$x \geq -\dfrac{10}{3}$ \qquad and \qquad $x \leq 2$

3. $x > -\dfrac{15}{2}$ and $x < \dfrac{9}{2}$

    (solve the two inequalities $2x + 3 < 12$ and $-(2x+3) < 12$)

$-\dfrac{15}{2}$ \qquad\qquad $\dfrac{9}{2}$

$x > -\dfrac{15}{2}$ \qquad and \qquad $x < \dfrac{9}{2}$

4. $x < \dfrac{4}{3}$ or $x > \dfrac{14}{3}$

(solve the two inequalities $3x - 9 > 5$ and $-(3x - 9) > 5$)

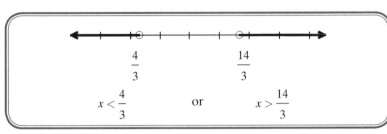

5

Relations, Functions, and Graphs

By now you have discovered that there are lots of rules in algebra. There are rules for how to add, subtract, multiply, and divide real numbers, and there are rules for how to multiply exponential expressions that have the same base.

Algebra also involves **expressions**. Algebraic expressions can be thought of as a set of instructions that are specific to the problem being considered. Different sets of instructions have different properties, and mathematicians have decided to categorize them as either relations or functions, depending on the instructions involved.

Lesson 5-1: Relations

A **relation** is a mathematical way of establishing a relationship between two quantities. One of the quantities is usually called the **input** and the other quantity is called the **output**. The input is also called the **domain,** and the output is often referred to as the **range** or the **co-domain**. Sometimes relations are written as a collection of ordered pairs.

5 **RELATIONS, FUNCTIONS, AND GRAPHS**

Ordered pairs are an efficient way to keep track of information. An **ordered pair** consists of two elements, the first element coming from the domain and the second one coming from the range. The two elements are written side by side, separated by a comma, and surrounded by parentheses. I have just used a lot of words to describe an object that looks something like (a, b) where a is the input and b represents the output corresponding to the input value a.

A relation can be viewed as a collection of ordered pairs. Relations are fairly relaxed in the sense that there are no restrictions or regulations regarding the number of output values that can be associated with a particular input value. An example of a relation is the collection of ordered pairs $(1, 1)$, $(1, 2)$, and $(2, 1)$. In this collection of ordered pairs, the input value 1 has two different output values (1 and 2). A relation can consist of *any* collection of ordered pairs.

Relations can be described in a variety of ways. You can describe a relation using words, a formula, a table of input and output values, a list of ordered pairs, or a graph.

It may be helpful to give you a couple of examples of relations. A list of ordered pairs may look something like $(1, 3)$, $(2, 6)$, $(2, 4)$, $(3, 9)$. This same relation could be represented using the table that follows.

Input	Output
1	3
2	6
2	4
3	9

The problem with tables and lists of ordered pairs is that you are limited to the data provided. In the relation described by the collection of ordered pairs $(1, 3)$, $(2, 6)$, $(2, 4)$, $(3, 9)$, you would not be able to determine the output value that corresponds to an input value of 5.

Not only that, but the output isn't necessarily a unique value determined by the input. As you can see in the previous table, an input of 2 results in an output of 6 *and* an output of 4. The fact that a particular input can result in two different outputs is not a good thing. Imagine taking a multiple choice test where there is more than one right answer and the answer that will be graded as "correct" depends on the mood of the teacher? Being able to have more than one output for a given input does not necessarily work to your advantage.

Relations can also be specified by writing a formula and using variables to represent the input and the output. A formula gives a set of specific instructions about what to do with the input value. If x represents the input and y represents the output of our relation, I could give you the formula $y = 2x + 1$ and have you calculate the output for several values of the input. Having a formula enables you to completely describe a relation without having to list every single ordered pair in the relation.

Relations that are described using words are very important in applying mathematics to our everyday lives. For example, suppose you have a job that pays $8 an hour for up to 40 hours and time-and-half for all hours over 40 hours in a week. You could use this relationship between time worked and money earned to calculate your paycheck each week.

Finally, you could be given the graph of a relation. A graph of a function helps you visualize the function so that you can understand its properties more thoroughly. For example, you could look at a graph of the price of a stock over time and decide whether to buy or sell shares of the stock.

Lesson 5-2: Functions

Like a relation, a function is also a set of instructions that establishes a relationship between two quantities. Functions have input and output values. The input is still called the domain and the output is still called the range or the co-domain. The variable used to describe the elements in the domain is called the **independent variable**, and the variable used to describe the elements in the range is called the

Relations, Functions, and Graphs

5

dependent variable. But there's more to a function than there is to a relation. With a relation the same input value could result in two or more different output values. With a function, this is not possible. The distinguishing feature of a function compared to a relation is that with a function, each input value is assigned a unique output value. We tend to focus on functions more than relations because of this extra stipulation.

A function is usually given a name. Sometimes a function is given the name y, other times it is given the name $f(x)$. It doesn't matter what you name a function, what matters is what you do with it.

Just like with relations, you can represent a function in a variety of ways. You can use words, a formula, a table of input and output values, a list of ordered pairs, or a graph to describe a function.

Lesson 5-3: Formulas

A function is a set of instructions telling you how to change the input into the output. A formula gives a very clear description of what the function does. It is a way to describe the function in mathematical terms instead of using words. For example, the function that takes the input and triples it and then adds 5 could be described by the formula $f(x) = 3x + 5$. The order of operations is very important when working with formulas. Multiplication comes before addition, so when you read $3x + 5$ you should realize that you need to take x and triple it first, then add 5. When you see a function like $f(x) = 3x + 5$ you should recognize the important features: the variable that appears in parentheses in the function name is the independent variable, and the formula for the function shows you how to transform the input. This formula instructs you to take whatever is in parentheses and triple it first, then add 5. Technically it doesn't really matter what is in parentheses. For example, $f(2) = 3 \cdot 2 + 5 = 11$ and $f(4) = 3 \cdot 4 + 5 = 17$. But in a similar manner, $f(\Theta) = 3 \cdot \Theta + 5$, and $f(2x) = 3(2x) + 5 = 6x + 5$. As long as you follow the rules and replace the independent variable with the object in parentheses, nothing can go wrong.

A useful skill to develop is the ability to describe a function if you are given a formula, and being able to write a formula if you are given its description. We will practice those skills in the next two examples.

Example 1

Write a formula to represent the function that first adds 10 to a number and then doubles the result.

Solution: Since the function first adds 10, we would want to start with $(x+10)$. Then we have to double this new number, so our formula would be $f(x)=2(x+10)$.

Example 2

Describe the following function in words: $f(x)=\dfrac{x-5}{4}$

Solution: This formula is a bit tricky to interpret, mainly because there are some invisible parentheses. The formula should actually be thought of as being $f(x)=\dfrac{(x-5)}{4}$. The formula takes a number and first subtracts 5. Then it divides the result by 4.

Lesson 5-3 Practice

1. Describe the formula in words: $f(x)=x^2$

2. Write a formula to represent a function that takes a number and divides it by 3, and then adds 12 to the result.

Lesson 5-4: Tables

Scientists often collect data from various instruments and recording that information in a table. These tables can represent functions, and provide an easy way to describe a complex formula. Reading a table is straightforward. If a table is written vertically then the column on the left represents the independent variable, or the input, and the column on the right represents the dependent variable, or the output.

The columns will be labeled, so if you are in doubt just look at the top row to see what each column represents. For example, the function described in the table below does not have a formula associated with it, but by reading the table it is possible to determine that $f(3)=7$ and $f(9)=12$.

x	$f(x)$
3	7
9	12

If a table is written horizontally then the top row usually represents the independent variable and the bottom row represents the dependent variable. But, if in doubt just check the first column to see what each row represents. For example, the function defined by the table shown here also does not have a formula associated with it, but you can see that $g(0)=-5$ and $g(1)=2$.

x	0	1
$g(x)$	–5	2

Lesson 5-4 Practice

Using the table below, determine $f(5)$ and $f(15)$.

x	$f(x)$
5	8
10	15
15	22
20	5

Lesson 5-5: The Cartesian Coordinate System

In order to graph a function you need to be able to graph both the input and the output values at the same time. You will need to use one number line for the input and another number line for the output.

If you arrange these two number lines so that they are perpendicular to each other, as is shown in Figure 5.1, you will have created what is known as the **Cartesian coordinate system**.

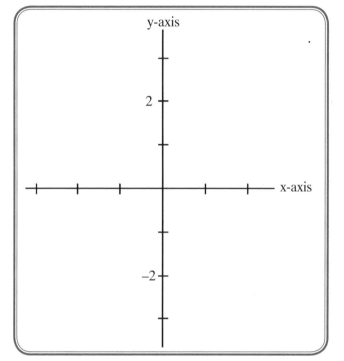

Figure 5.1: The Cartesian coordinate system.

The Cartesian coordinate system is named for Rene Descartes, who is credited with inventing this system. This coordinate system brings geometry and algebra together, enabling us to use algebra to solve problems in geometry and to use geometry to gain insight into algebraic results. We use this system, which is sometimes called the coordinate plane, to locate points and draw figures.

The horizontal number line is called the **x-axis** and it is used to record the values of the input, or independent variable. The vertical number line is called the **y-axis** and it is used to record the function values, or the output. The two lines intersect at 0; the point where the two lines intersect is called the origin.

5
RELATIONS, FUNCTIONS, AND GRAPHS

Two numbers are used to describe the location of a point in the plane, and they are recorded in the form of an ordered pair (x, y), where the first number represents the horizontal distance (using a horizontal number line) from the y-axis to the point and the second number represents the vertical distance (using a vertical number line) from the x-axis to the point. This should look familiar to you; recall that relations and functions are sometimes given as a list of ordered pairs. The first coordinate of the ordered pair is called the **x-coordinate** and the second coordinate is called the **y-coordinate**. Points that are on the x-axis have a y-coordinate equal to 0, and points that are on the y-axis have their x-coordinate equal to 0. Those points that lie to the right of the y-axis have a positive x-coordinate and points to the left of the y-axis have a negative x-coordinate. Similarly, points above the x-axis have a positive y-coordinate and points below the x-axis have negative a y-coordinate. The points (2, 3) and (–1, 2) are shown in Figure 5.2.

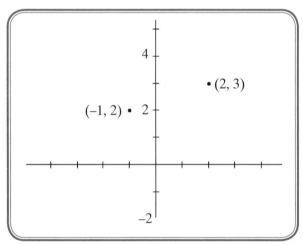

Figure 5.2: The graph of the points (2, 3) and (–1, 2).

The coordinate axes divide the plane into four parts, called quadrants. Quadrant I consists of those points that have positive values for both their x-coordinate and y-coordinate. Quadrant I is located in the upper right part of the plane. We continue on to Quadrants II, III, and IV in a counter-clockwise progression. Quadrant II consists of those points that have a negative value for their x-coordinate and a positive value for their y-coordinate. Quadrant III consists of those points that have negative values for both their x-coordinate and their y-coordinate. Finally, Quadrant IV consists of those points that have a positive value for their x-coordinate and a negative value for their y-coordinate. The signs for the x-coordinates and y-coordinates are summarized in the table shown here.

Quadrant	Coordinate Signs
I	(+, +)
II	(−, +)
III	(−, −)
IV	(+, −)

Lesson 5-5 Practice

1. Graph points with coordinates (−3, −1) and (2, −4).

2. Read off the coordinates of the points P and Q shown in Figure 5.3.

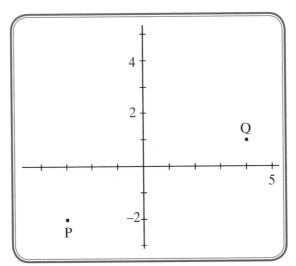

Figure 5.3: Points P and Q.

5
RELATIONS, FUNCTIONS, AND GRAPHS

Lesson 5-6: Graphs of Functions and Relations

It is in our nature to look for patterns and relationships between objects or concepts. A picture certainly helps us discover these patterns, which is why graphing functions and relations is so important. Graphing a function or a relation can involve plotting a lot of points in the coordinate plane and then trying to connect the dots. But, if you can plot one point, you can plot them all, and if you plot enough points you will gain some insight into the function or relation you are trying to study.

In this section, we will focus on graphing functions and relations that are specified by a list of ordered pairs, a table, and a formula. We will explore graphs in the next few chapters as well, and even after that we will have only touched the surface of graphing. A graphing calculator can be used to graph many complicated formulas, and still there will be more to learn about graphing. It truly is a rich part of mathematics.

A graph of a list of ordered pairs is sometimes called a **scatter plot**. To graph a list of ordered pairs, just graph each ordered pair on the list. This is also a good time to review the differences between relations and functions, and to get some graphical insight into their differences.

Example 1

Graph the following set of ordered pairs:
$\{(1,1),(2,3),(1,0),(0,2)\}$ and determine if it is a relation or a function.

Solution:

The graph is shown in Figure 5.4. It is a relation and not a function because the input value of 1 has two different output values.

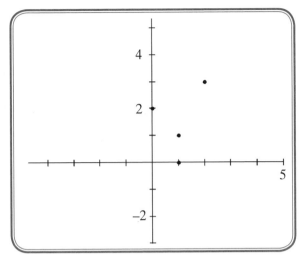

Figure 5.4: The graph of $\{(1,1),(2,3),(1,0),(0,2)\}$.

Example 2

Graph the following and determine if it is a relation or a function.

x	–2	0	2
$f(x)$	1	3	5

Solution:

The graph is shown in Figure 5.5. It is a function because each input value has only one unique output value.

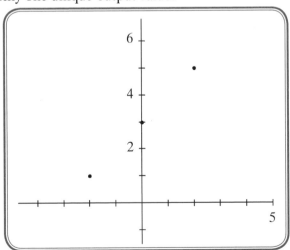

Figure 5.5: The graph of the function defined in Example 2.

RELATIONS, FUNCTIONS, AND GRAPHS

5

Example 3

Graph the function $f(x)=x^2$ for the domain $\{-1,0,1\}$.

Solution:

We need to generate the ordered pairs to graph. Plug each number in the domain into the formula for $f(x)$ to find the corresponding ordered pair. Then graph the ordered pairs. The domain $\{-1,0,1\}$ will generate the ordered pairs (–1, 1), (0, 0), and (1, 1). The graph of these ordered pairs is shown in Figure 5.6. It is a function because each input value has only one unique output value.

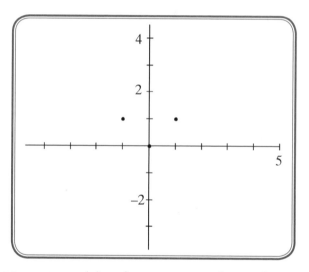

Figure 5.6: The graph of $f(x)=x^2$ for the domain $\{-1,0,1\}$.

Example 4

Graph the following: $\{(0,2),(1,2),(2,2),(3,2)\}$ and determine if it is a relation or a function.

Solution:

The graph is shown in Figure 5.7. It is a function because each input value has only one output value.

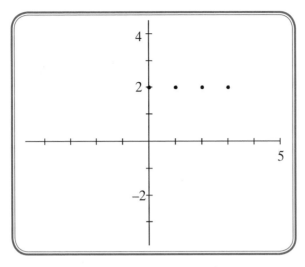

Figure 5.7: The graph of $\{(0,2),(1,2),(2,2),(3,2)\}$.

Lesson 5-6 Practice

Graph the following and determine if they are relations or functions:

1. $\{(0,2),(1,3),(2,2),(1,0)\}$.

2.

x	$f(x)$
-2	1
-1	0
0	-1

3. $f(x)=2x$ for the domain $\{-1,0,1,2\}$.

Answer Key
Lesson 5-1

1. $f(x)=x^2$: the function takes a number and squares it.

2. $f(x)=\dfrac{x}{3}+12$

Lesson 5-4

1. $f(5)=8,\ f(15)=22$

Lesson 5-5

1. Graph of the points $(-3, -1)$ and $(2, -4)$.

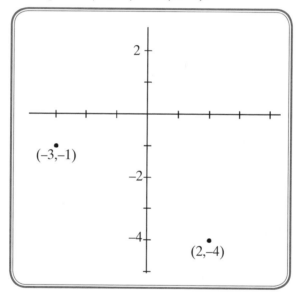

2. P is the point $(-3, -2)$ and Q is the point $(4, 1)$.

Lesson 5-6

1. Graph of the points $\{(0,2),(1,3),(2,2),(1,0)\}$; it is a relation.

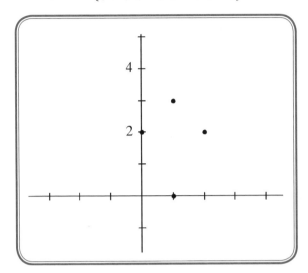

2. Graph of the points $\{(-2,1),(-1,0),(0,-1)\}$; it is a function.

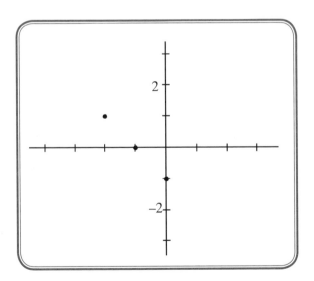

3. Graph of the points $\{(-1,-2),(0,0),(1,2),(2,4)\}$; it is a function.

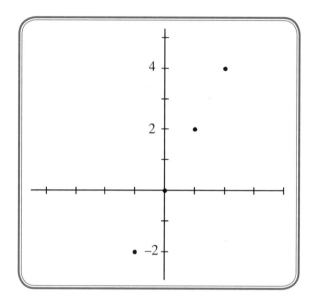

6

Linear Equations

A linear function is one of the nicest functions to work with. A linear function involves two constants and one variable raised to the first power. The function $f(x) = 3x + 2$ is an example of a linear function. Sometimes the function is called y instead of $f(x)$; you may see the function $f(x) = 3x + 2$ written as $y = 3x + 2$. It doesn't matter what we choose to name the function, what matters is the formula that describes the function. The functions $y = 3x + 2$ and $f(x) = 3x + 2$ have different names but are described using the same formula, so these two functions are really the same. We can examine this function in more detail in the next section and make some observations. We will then turn our observations into generalizations about all linear functions.

Lesson 6-1: Linear Functions and Calculating Slopes

The best way to understand a function is to generate a few points and plot them. Whenever you are trying to examine a function it is best to

6

LINEAR EQUATIONS

x	$f(x)=3x+2$
−3	−7
−2	−4
−1	−1
0	2
1	5
2	8
3	11

pick points in a systematic way. I recommend using a combination of evenly spaced negative and positive values for x. Evaluating the function when $x = 0$ is also helpful. I've evaluated the function $f(x)=3x+2$ for seven values of x and given the corresponding function values in the table below. Notice that all of the values of x are evenly spaced and center around $x = 0$. Remember that x is called the independent variable. By "independent," we mean that we have control over what values we use (within reason, of course). The function $f(x)$ or the variable y is called the dependent variable. By "dependent" we mean that the value of the function depends on what value x takes on.

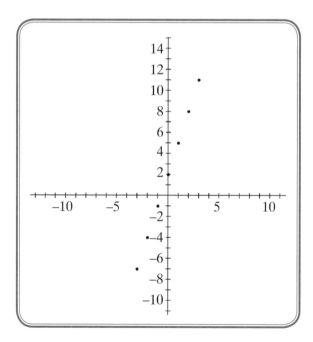

Figure 6.1: The graph of $f(x)=3x+2$.

Once you have determined some of the ordered pairs that are generated by the function, I recommend graphing them. A carefully drawn picture or a graph can give lots of insight into the nature of the function. The graph of these ordered pairs is shown in Figure 6.1 on page 122.

Now that we have a picture of this function we can make some observations. First of all, notice that all of the points appear to lie on the same line. This function was called a linear function, and now that makes sense. Linear functions with one independent variable and one dependent variable have graphs that look like lines.

The graph of $f(x) = 3x + 2$ is a line that is not horizontal like the x-axis, nor is it vertical like the y-axis. It is worth measuring the slant, or slope, of lines that are not vertical or horizontal. The slope of a line is defined as the ratio of the "rise" of a graph divided by the "run" of the graph, or the "rise over the run" of the graph. The rise of the graph is a measurement of the change in the dependent variable (the variable y), and the run of the graph is a measurement of the change in the independent variable (the variable x):

$$\text{slope} = \frac{\text{change in } y}{\text{change in } x} = \frac{\Delta y}{\Delta x}$$

Mathematicians often use the symbol Δ to represent change. In geometry we use the symbol Δ to represent a triangle, but in algebra it is a short-hand way to abbreviate "change." The change in a variable is just the final value of the variable minus the initial value of the variable. You can calculate the slope of a line using any two points on the line. If you have two points, say (a, b) and (c, d), then the change in dependent variable would be found by taking the difference between the y-coordinates of the two points: $\Delta y = d - b$. The change in independent variable would be found by taking the difference between the x-coordinates of the two points: $\Delta x = c - a$. The slope of the line is found by using the previous equation:

$$\text{slope} = \frac{\text{change in } y}{\text{change in } x} = \frac{\Delta y}{\Delta x} = \frac{d-b}{c-a}$$

You may be wondering how I knew that the point (c, d) was the final point and (a, b) was the initial point. In reality, it doesn't matter. If I had switched them so that (a, b) was the final point and (c, d) was the initial point, the result would have been the same:

$$\text{slope} = \frac{\text{change in } y}{\text{change in } x} = \frac{\Delta y}{\Delta x} = \frac{b-d}{a-c} = \frac{(-1)(d-b)}{(-1)(c-a)} = \frac{d-b}{c-a}$$

It is not important which point is considered to be the final point and which point is considered to be the initial point. It is important that the change in y is put into the numerator of the ratio and the change in x is put into the denominator of the ratio. The other thing that is important is that you are *consistent* in which point is the final point and which point is the initial point when calculating the changes. If (a, b) is the final point when calculating Δy, it must also be the final point when calculating Δx. Let's use a couple of different points in the table to calculate the slope of the line $y = 3x + 2$:

Point Number	x	$f(x) = 3x + 2$
1	−3	−7
2	−2	−4
3	−1	−1
4	0	2
5	1	5
6	2	8
7	3	11

Example 1

Calculate the slope of $f(x) = 3x + 2$ using point #1 and point #3.

Solution:

Point #1 is the point (–3, –7) and point #3 is the point (–1, –1).
The slope is:

$$\text{slope} = = \frac{\Delta y}{\Delta x} = \frac{-1 - (-7)}{-1 - (-3)} = \frac{-1 + 7}{-1 + 3} = \frac{6}{2} = 3.$$

Example 2

Calculate the slope of $f(x) = 3x + 2$ using point #4 and point #7.

Solution:

Point #4 is the point (0, 2) and point #7 is the point (3, 11).
The slope is:

$$\text{slope} = \frac{\Delta y}{\Delta x} = \frac{11 - 2}{3 - 0} = \frac{9}{3} = 3$$

We calculated the same slope in Example 1 and in Example 2. In fact, one of the things that make lines so special is that their slope is *constant*. It doesn't matter which set of points you use to calculate the slope of a line, the slope will always be the same for a particular line.

Notice that the slope of the linear function $f(x) = 3x + 2$ is 3; 3 also happens to be the coefficient in front of x. Could this be a coincidence? No! It is easy to pick out the slope of a linear function if the function is written in this particular form: read off the coefficient in front of x. For example, the slope of the linear function $f(x) = 7x - 9$ is...7!

Example 3

Pick 2 points and calculate the slope of the function:
$$f(x) = -2x + 1$$

Solution:

We can see by the formula for $f(x)$ that the slope is –2. It doesn't matter which two points I pick, so I will pick points corresponding to $x = 0$ and $x = 1$. The point corresponding to $x = 0$ is (0, 1) (since $f(0)=1$) and the point corresponding to $x = 1$ is (1, –1) (since $f(1)=(-2)\cdot1+1=-1$). The slope of the line passing through (0, 1) and (1, –1) is:

$$slope==\frac{\Delta y}{\Delta x}=\frac{-1-1}{1-0}=\frac{-2}{1}=-2$$

Example 4

Find the slope of the line that passes through the points (4, –2) and (6, 3).

Solution:

Use the equation for the slope:

$$slope==\frac{\Delta y}{\Delta x}=\frac{-2-3}{4-6}=\frac{-5}{-2}=\frac{5}{2}$$

Lesson 6-1 Practice

1. Fill in the table and then graph the points for the function $f(x)=-x+3$.

x	$f(x)=-x+3$
–2	
–1	
0	
1	
2	

2. Find the slope of the function $f(x)=2x-4$.

3. Find the slope of the line that passes through the points (–2, 1) and (4, –3).

Lesson 6-2: Intercepts

In addition to the slope of a line, there are other important features of a line that are worth mentioning. Two important points of a line are called the intercepts. There is an **x-intercept**, which is where the graph of the line crosses the x-axis, and there is a **y-intercept**, which is where the graph of the line crosses the y-axis. Finding them is fairly straightforward.

The y-intercept of a function is where that function intersects the y-axis. The y-axis consists of all points in the plane where $x = 0$. So, to find the y-intercept of a function you need to evaluate the function at $x = 0$. The y-intercept is the point $(0, f(0))$.

Example 1

Find the y-intercept of the following functions:

a. $f(x) = 3x + 1$

b. $f(x) = -2x - 3$.

Solution:

a. The y-intercept is the point $(0, f(0))$: if $f(x) = 3x + 1$, then $f(0) = 3 \cdot 0 + 1 = 1$. The y-intercept is the point (0, 1).

b. The y-intercept is the point $(0, f(0))$: if $f(x) = -2x - 3$, then $f(0) = -2 \cdot 0 - 3 = -3$. The y-intercept is the point (0, –3).

Finding the x-intercept of a line requires a little more work than does finding the y-intercept. Functions are written so that the values of y are easy to determine, but the price is that the values of x are not as easy to find. The x-intercept is the point where the line crosses the x-axis; the x-axis is the set of all points with a y-coordinate equal to 0. To find the x-intercept of a line, set the function $f(x)$ equal to 0 and solve for x. Because functions are sometimes called $f(x)$ and other times called y, finding the x-intercept involves setting y equal to 0 and solving for x.

Example 2

Find the x-intercept of the following functions:

a. $f(x)=3x+1$

b. $f(x)=-2x-3$

Solution:

a. Set $f(x)=0$ and solve for x:

$$0=3x+1$$
$$-3x=1$$
$$x=-\frac{1}{3}$$

The x-intercept is the point $\left(-\frac{1}{3},0\right)$.

b. Set $f(x)=0$ and solve for x:

$$0=-2x-3$$
$$2x=-3$$
$$x=-\frac{3}{2}$$

The x-intercept is the point $\left(-\frac{3}{2},0\right)$.

Notice that the y-intercept can be directly read off of the equation, but finding the x-intercept requires us to solve an algebraic equation.

Lesson 6-2 Practice

Find the slope, the x-intercept, and the y-intercept of the following lines:

1. $f(x)=3x+2$

2. $f(x)=5x+10$

3. $y=-2x-4$

Lesson 6-3: Point-Slope Form

The fact that the slope of a line is always the same, regardless of which two points on the line you use to calculate it, enables us to find the equation of a line very easily. If you need to find the equation of a line, all you need are two things: a point and the slope. Think of a line as a set of points. Suppose you know the slope of a line and also one point on the line. Then knowing the equation of the line will enable you to find the corresponding y-coordinate for any given x-coordinate. I will discuss several methods to determine the equation of a line, and there are many forms we use to write the equation of a line. The first method involves using the point-slope formula, and it makes use of the constant slope of a line.

Suppose that you know that a line passes through the point (a, b) and it has a slope m. If (x, y) is any other point on the line, then the slope between (a, b) and (x, y) must be m. Let's calculate the slope between (a, b) and (x, y):

$$\text{slope} = \frac{\Delta y}{\Delta x} = \frac{y - a}{x - b} = m$$

We can then multiply both sides of this equation by $(x - b)$:

$$\cancel{(x - b)} \cdot \frac{(y - a)}{\cancel{(x - b)}} = (x - b) \cdot m$$

$$y - a = m(x - b)$$

This last equation is the **point-slope equation** for a line. If you know one point on the line, (a, b), as well as the slope, m, you can use this equation to write the equation of the line.

Example 1

Find the point-slope form of the equation of the line passing through the point (1, 3) having slope 5.

Solution:

Using the point-slope formula, the equation of the line is
$y - 3 = 5(x - 1)$.

6

LINEAR EQUATIONS

Example 2

Find the point-slope form of the equation of the line passing through the points (–1, 4) and (3, 2).

Solution: The point-slope formula requires us to know a point and a slope. We are given two points, so the first thing we need to find is the slope:

$$\text{slope} = = \frac{\Delta y}{\Delta x} = \frac{2-4}{3-(-1)} = \frac{-2}{4} = -\frac{1}{2}$$

Now that we have the slope we are ready to use the point-slope equation. It doesn't matter which point we use.

I will use (3, 2): $y - 2 = -\frac{1}{2}(x - 3)$

This equation is equivalent to the equation $y - 4 = -\frac{1}{2}(x - (-1))$, which is the equation you would have generated if you used the point (–1, 4). If you solve each equation for y, you will see that you have the same formula in both cases, which is why the two representations are equivalent.

The point-slope form of a line is usually an intermediate step. We usually like our equations to be in slope-intercept form or in standard form, which we will discuss next.

Lesson 6-3 Practice

Use the point-slope formula to find the equation of the following lines:

1. The line passing through the point (1, 2) having slope 3.

2. The line passing through the points (2, –1) and (4, 1).

3. The line passing through the points (–4, 2) and (0, 1).

Lesson 6-4: Slope-Intercept Form

The **slope-intercept form** of the equation of a line is written

$$y = mx + b$$

where m is the slope and b is the y-intercept. This form should look familiar: when I discussed linear functions earlier I used functions of the form $f(x) = mx + b$. A function by any other name (like y) represents the same thing; you already have experience using the slope-intercept form of a line. It is not surprising that the slope-intercept form of a line is sometimes called the **function form** of a line. Now it is time to practice writing equations of lines in this form. There are several ways to approach these problems. If you are given the slope and the y-intercept directly, just put that information into the slope-intercept equation of a line. Otherwise, I recommend starting with the point-slope equation of a line and then solving for y.

Example 1

Find the slope-intercept equation of the line with slope 2 and y-intercept –3.

Solution: The slope-intercept equation of a line is $y = mx + b$. We are given both the slope and the y-intercept, so just plug them into the equation. The equation of the line is then $y = 2x - 3$.

Example 2

Find the slope-intercept equation of the line passing through (–2, 5) having slope 4.

Solution: In this case we are given a point, (–2, 5), and the slope, 4. The slope-intercept equation of a line has y on one side of the equal sign and everything else on the other side. To find the slope-intercept equation of this line we need to start with the point-slope equation of a line and then solve for y:

$$y - 5 = 4(x - (-2))$$

Start with the point-slope form of a line: $y - 5 = 4(x + 2)$

Distribute the 4: $y - 5 = 4x + 8$

Add 5 to both sides: $y = 4x + 13$

Example 3

Find the slope-intercept equation of the line passing through (3, 4) and (–1, 1).

Solution: In order to use the point-slope equation we need a point, which we have, and a slope, which we don't. So first find the slope:

$$slope = = \frac{\Delta y}{\Delta x} = \frac{4-1}{3-(-1)} = \frac{3}{4}$$

Next, use the point-slope formula and solve for y (it doesn't matter which point you use):

$$y - 4 = \frac{3}{4}(x-3)$$

Start with the point-slope form of a line and distribute the $\frac{3}{4}$:

$$y - 4 = \frac{3}{4}x - \frac{9}{4}$$

Add 4 to both sides:

$$y = \frac{3}{4}x - \frac{9}{4} + 4$$

Get a common denominator to add the fractions:

$$y = \frac{3}{4}x - \frac{9}{4} + \frac{16}{4}$$

Simplify:

$$y = \frac{3}{4}x + \frac{7}{4}$$

Lesson 6-4 Practice

Find the slope-intercept equation of the following lines:

1. The line with slope 6 and y-intercept –2.

2. The line that passes through the point (1, 4) and has slope –3.

3. The line that passes through the points (2, –3) and (–2, 6).

Lesson 6-5: Standard Form

The equation of a line written in **standard form** looks like

$$Ax + By = C$$

where A, and B represent constants that are in front of the variables and C is a constant. The variables x and y represent the independent and dependent variables, respectively. There are many times that you will be given an equation in standard form and will have to extract useful information from it. Other times you will be asked to write your equation in standard form with integer coefficients. If you are given the equation in standard form, often times the best thing to do is to put it in slope-intercept form. Then you will be able to read off the slope and the y-intercept easily. What you do depends on the instructions given in the problem. So be sure to read each problem carefully so you can answer it correctly.

Example 1

Write the equation $y = \frac{2}{5}x + 2$ in standard form with integer coefficients.

Solution:
To write the equation in standard form with integer coefficients, the first thing you need to do is clear out all of the fractions. You'll want to carefully multiply both sides of the equation by 5 (using the distributive property where necessary) so that all of the coefficients are integers. Then you need to make sure that all of the variables are on one side, numbers on the other:

$$y = \frac{2}{5}x + 2$$
$$5 \cdot y = 5 \cdot \left(\frac{2}{5}x + 2\right)$$
$$5y = 2x + 10$$
$$-2x + 5y = 10$$

Example 2

Find the equation of the line in standard form that has slope –2 and passes through (1, 4).

Solution:

Start with the point-slope equation, and simplify, moving the variables to one side and the numbers to the other:

$$y - 4 = -2(x - 1)$$
$$y - 4 = -2x + 2$$
$$y = -2x + 6$$
$$2x + y = 6$$

Example 3

Find the equation of the line in standard form with integer coefficients that passes through (–3, 1) and (2, 2).

Solution:

In order to find the equation of the line we need a point and a slope. So we need to find the slope:

$$\text{slope} = = \frac{\Delta y}{\Delta x} = \frac{2 - 1}{2 - (-3)} = \frac{1}{5}$$

Now we are ready to use the point-slope formula, clear out the fractions and simplify:

$$y - 2 = \frac{1}{5}(x - 2)$$
$$5 \cdot (y - 2) = 5 \cdot \frac{1}{5}(x - 2)$$
$$5y - 10 = x - 2$$
$$5y = x + 8$$
$$-x + 5y = 8$$

It is worthwhile to examine the relationship between standard form and slope-intercept form. If we start with an equation in standard form and solve for y, our line will then be expressed in slope-intercept form:

$$Ax + By = C$$

$$By = -Ax + C$$

$$y = -\frac{A}{B}x + \frac{C}{B}$$

So if you have an equation in standard form $Ax + By = C$, then the slope is $-\frac{A}{B}$ and the y-intercept is $\frac{C}{B}$. There are some people who would just memorize these formulas, but if I remembered every formula I was ever shown, I would never be able to remember my phone number! I prefer to understand where the formulas come from. That way I don't have to rely on my memorization skills. All you have to realize is that if you have an equation in standard form, the key to determining the slope and the y-intercept is to solve for y!

Lesson 6-5 Practice

Find the equation of the following lines. Write your answer in standard form with integer coefficients.

1. The line with slope –4 and y-intercept $\frac{1}{2}$.

2. The line passing through the point (–1, 4) having slope 2.

3. The line passing through the points (3, –2) and (5, 2).

Lesson 6-6: Horizontal and Vertical Lines

Lines can be horizontal, vertical, inclined to the left, or inclined to the right. As we have already discussed, lines that are inclined to the left or the right have slopes that can be calculated using the equation

$$\text{slope} = \frac{\Delta y}{\Delta x}$$

Horizontal and vertical lines are special lines with special properties. Horizontal and vertical lines are in a class by themselves, so it's best to examine them separately.

6

LINEAR EQUATIONS

Any line that is parallel to the x-axis is a **horizontal line**. The x-axis itself is an example of a horizontal line. What distinguishes a horizontal line from every other kind of line is that the y-coordinate of every point on a horizontal line *is the same*, and a horizontal line can be completely specified by specifying what that y-coordinate is. Every horizontal line can be written as $y = a$, where a is a real number representing the y-coordinate of that line. For example, the equation of the x-axis is $y = 0$.

Any line that passes through two points that have the same y-coordinate, but different x-coordinates, will be horizontal. We can calculate the slope of a horizontal line using the equation for the slope of a line:

$$\text{slope} = \frac{\Delta y}{\Delta x}$$

Since the y-coordinates of all points on a horizontal line have the same value, the rise of a horizontal line is always 0, making the slope of a horizontal line 0 (since 0 divided by any non-zero number is 0). A horizontal line has a whole lot of *run* and no *rise*. An important characteristic of a horizontal line is that its slope is 0. That's not surprising because a horizontal line has the equation $y = a$ and we can re-write the equation $y = a$ as $y = 0 \cdot x + a$.

A vertical line is any line that is parallel to the y-axis. With a vertical line, all of the points on the line have the same x-coordinate. Writing the equation of a vertical line involves specifying the x-coordinate, and in general it is written $x = a$ where a is a real number that represents the x-coordinate of the line. A vertical line has a whole lot of *rise* and no *run*. This means that $\Delta x = 0$. If you tried to calculate the slope of a vertical line you would run into problems because the denominator of the slope (the run) would be 0, and we aren't allowed to divide by 0. So there is no point in calculating the slope of a vertical line and we say that a vertical line has no slope. Recognize that there is a difference between "zero slope" (i.e. a horizontal line) and "no slope" (i.e.

a vertical line). "No" and 0 mean two different things in this context, and you must be careful to mean what you say and say what you mean.

The nice thing about vertical and horizontal lines is that all you need is 1 point to describe them, because you already know the slope (if the slope exists then the line is horizontal with slope 0 and if the slope doesn't exist then the line is vertical).

Example 1

Write the equation of the horizontal line passing through (3, 2).

Solution: All you need to describe a horizontal line is the y-coordinate of a point. The equation of the horizontal line passing through (3, 2) is $y = 2$.

Example 2

Write the equation of the vertical line passing through (4, –2).

Solution: All you need to describe a vertical line is the x-coordinate of a point. The equation of the vertical line passing through (4, –2) is $x = 4$.

Example 3

Find the equation of the line passing through (2, –1) and (2, 3).

Solution: Because the x-coordinates of these two points are the same, we are dealing with the vertical line $x = 2$.

Example 4

Find the equation of the line passing through (–2, 1) and (3, 1).

Solution: Because the y-coordinates of these two points are the same, we are dealing with the horizontal line $y = 1$.

6

LINEAR EQUATIONS

Lesson 6-6 Practice

Find the equation of the following lines.

1. The vertical line that passes through the point (1, –3).

2. The horizontal line that passes through the point (–4, 7).

Lesson 6-7: Graphing Linear Equations

Now that we understand the various ways to write an equation of a line, it's time to get a picture of what our lines look like. The key to graphing linear functions, or lines, is that all you need is two points and a straightedge (or ruler).

In order to find two points on the line, pick two different values for x, substitute those values into the equation to find the corresponding values for y. Then plot the two ordered pairs and connect the dots with a ruler. That's all there is to it. You can use any two points that you want.

Another way to graph a line is to find both intercepts and plot them. The way to find the x-intercept is to set $y = 0$ and solve for x; to find the y-intercept set $x = 0$ and solve for y. I'll use the intercept method in my examples, just so that we can review how to find them as well as how to solve equations. The more you practice, the better you'll get. You don't have to do it my way, though. You can draw your graphs using your two favorite points and compare your graphs to mine. It doesn't matter which two points you plot and connect, as long as the points actually lie on the line in question.

Example 1

Graph the line given by the equation: $y = 2x + 1$

Solution: First, find the y-intercept by setting $x = 0$:

$y = 2x + 1$

$y = 2 \cdot 0 + 1 = 1$

Then find the x-intercept by setting $y = 0$:

$y = 2x + 1$

$0 = 2x + 1$

$-2x = 1$

$x = -\dfrac{1}{2}$

Then plot the two intercepts $(0, 1)$ and $\left(-\dfrac{1}{2}, 0\right)$ and connect the points.

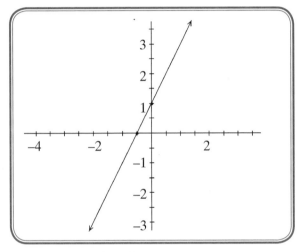

Figure 6.2: The graph of $y = 2x + 1$.

Example 2

Graph the line given by the equation: $3x + 2y = 6$

Solution: First, find the y-intercept by setting $x = 0$:

$3x + 2y = 6$

$3 \cdot 0 + 2y = 6$

$2y = 6$

$y = 3$

Then find the x-intercept by setting $y = 0$:

$3x + 2y = 6$

$3x + 2 \cdot 0 = 6$

$3x = 6$

$x = 2$

Then plot the two intercepts $(0, 3)$ and $(2, 0)$ and connect the points.

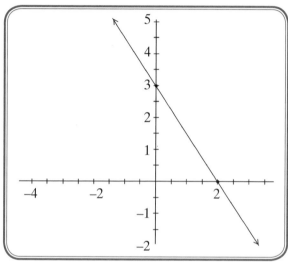

Figure 6.3: The graph of 3x + 2y = 6.

Example 3

Graph the line given by the equation $y = 3x$.

Solution:

First, find the y-intercept by setting $x = 0$: $y = 3 \cdot 0 = 0$. Notice that this line passes through the origin $(0, 0)$ which means that the x-intercept and the y-intercept are the same point. So we will need to pick a different value of x in order to come up with another point on the line. It doesn't matter what value we pick for x; to keep it simple, pick $x = 1$. Use that value to solve for y: $y = 3 \cdot 1 = 3$. Then plot the two points $(0, 0)$ and $(1, 3)$ and connect them.

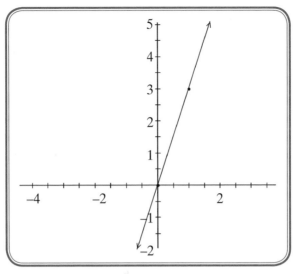

Figure 6.4: The graph of the line y = 3x.

Lesson 6-7 Practice

Graph the following lines:

1. $y = -2x + 1$

2. $x - 3y = 3$

3. $y = \dfrac{3}{2}x$

Answer Key
Lesson 6-1

1.

x	$f(x)=-x+3$
–2	5
–1	4
0	3
1	2
2	1

Graph of the points $\{(-2,5)(-1,4),(0,3),(1,2),(2,1)\}$

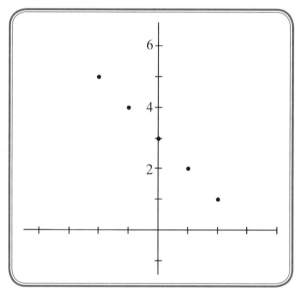

2. The slope is 2.

3. The slope is $-\dfrac{2}{3}$.

Lesson 6-2

1. The slope is 3, the x-intercept is $\left(-\dfrac{2}{3},0\right)$, and the y-intercept is $(0, 2)$.

2. The slope is 5,
   the x-intercept is (–2, 0), and
   the y-intercept is (0, 10).

3. The slope is –2,
   the x-intercept is (–2, 0), and
   the y-intercept is (0, –4).

Lesson 6-3

1. $y - 2 = 3(x - 1)$

2. $y - 1 = 1(x - 4)$ or $y + 1 = 1(x - 2)$, depending on the point used.

3. $y - 1 = -\frac{1}{4}(x - 0)$ or $y - 2 = -\frac{1}{4}(x + 4)$, depending on the point used.

Lesson 6-4

1. $y = 6x - 2$

2. $y = -3x - 7$

3. The slope is $-\frac{9}{4}$.
   Use the point-slope formula and solve for y: $y = -\frac{9}{4}x + \frac{3}{2}$

Lesson 6-5

1. $8x + 2y = 1$

2. $-2x + y = 6$

3. The slope is 2; use the point-slope formula and move the variables to the left of the equal sign: $-2x + y = -8$

Lesson 6-6

1. Read off the x-coordinate of the point: $x = 1$

2. Read off the y-coordinate of the point: $y = 7$

Lesson 6-7

1. Graph of the line $y = -2x + 1$
   (x-intercept $\left(\frac{1}{2}, 0\right)$, y-intercept $(0, 1)$)

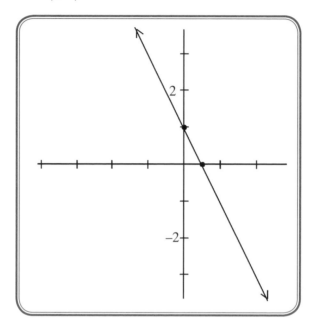

2. Graph of the line $x - 3y = 3$ (x-intercept $(3, 0)$, y-intercept $(0, -1)$).

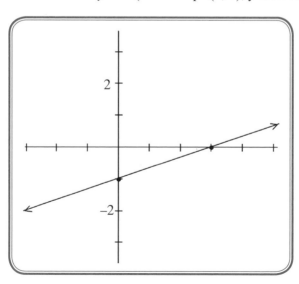

3. Graph of the line $y = \dfrac{3}{2}x$ (x-intercept and y-intercept are both (0, 0), find another point: (2, 3)

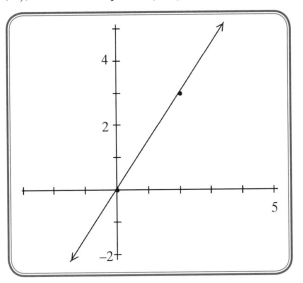

6 LINEAR EQUATIONS

7

Systems of Linear Equations

A **system of linear equations** refers to a *collection* of linear equations. A solution to a system of linear equations consists of all of the ordered pairs (a, b) that satisfy all of the equations in the system. That means that any solution of a system of linear equations must lie on all of the individual lines that make up the system. In other words, the solution of a system of equations consists of all points that lie at the intersection of all of the linear equations in the system.

I will discuss three ways to solve systems of linear equations. While these are not the only methods available, they are the ones that use the skills that you have learned so far.

When solving systems of linear equations, it is important to keep in mind how many points there are that satisfy all of the equations in the system. The key to understanding how many answers you are looking for is to keep in mind the properties of lines. Recall that two points determine the equation of a line. If you have two lines, there are only a few possibilities:

- The two lines are parallel and do not intersect. In this case, there are no points that lie on both lines. In other words, there is no solution to the system of equations.

- The two lines are different representations of the same line and the two equations are equivalent. If the two equations are equivalent then every point that satisfies one of the equations also satisfies the other equation. In other words, there are infinitely many solutions to the system of equations.

- The two lines are not parallel and the equations are not equivalent to each other. In this case, the two lines will intersect at one unique point and there is only one solution to the system of equations.

We will see examples of each of these cases, though our focus will be on the third situation.

Lesson 7-1: The Graphical Method

The most direct method for solving systems of equations is to graph all of the equations in that system and find the point of intersection. Because the solution of a linear system of equations must satisfy each equation in the system, the solution must lie on the graph of both equations. If the solution has integer values for both coordinates, a careful graph will yield the solution directly. Otherwise, the graphical solution will only give an approximate answer and indicate in which quadrant the solution lie. Any answer that you get using this method must be checked by substituting the values for x and y into both equations. If they are both satisfied then your answer is correct.

Example 1

Solve the system of equations given by: $\begin{cases} x + y = 4 \\ 3x - y = 0 \end{cases}$

Solution: Graph each line separately and look for the point of intersection. When graphing a line all you need is two points.

I usually pick the intercepts whenever possible. If a line goes through the origin, then both the x- and y-intercepts are the same point. In that case I abandon my bias toward the intercepts use another point on the graph as my second point. To find the x-intercept of the line $x + y = 4$ substitute $y = 0$ into the equation and solve for x:

$x + 0 = 4$

The x-intercept is the point $(4, 0)$. To find the y-intercept of the line $x + y = 4$ substitute $x = 0$ into the equation and solve for y:

$0 + y = 4$

The y-intercept is the point $(0, 4)$. Similarly, the x-intercept (and the y-intercept) of the line $3x - y = 0$ is the origin, or $(0, 0)$. I will need to find a second point, so I will let $x = 1$ and solve for y:

$3x - y = 0$

$3 \cdot 1 - y = 0$

$y = 3$

So the second point on the line $3x - y = 0$ is $(1, 3)$. The graph of both of these lines is shown in Figure 7.1. Looking at the graph we can see that the intersection of these two points is the point $(1, 3)$.

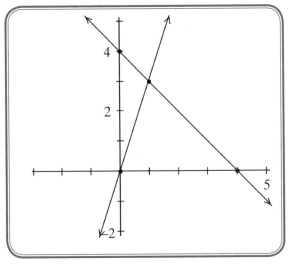

Figure 7.1: The graph of the system of equations.

$$\begin{cases} x+y=4 \\ 3x-y=0 \end{cases}$$

The last thing to do is check to see if the point (1, 3) satisfies both equations. Check the first equation:

$x + y = 4$

$1 + 3 = 4$

Check the second equation:

$3x - y = 0$

$3 \cdot 1 - 3 = 0$

The point (1, 3) satisfies both equations. Therefore the point (1, 3) is a solution of the system of equations $\begin{cases} x+y=4 \\ 3x-y=0 \end{cases}$

Lesson 7-1 Practice

Solve the following systems of equations by graphing each equation:

1. $\begin{cases} x-2y=0 \\ 2x+y=10 \end{cases}$
2. $\begin{cases} 2x+3y=13 \\ x-y=-1 \end{cases}$
3. $\begin{cases} x-y=-3 \\ x+y=1 \end{cases}$

Lesson 7-2: The Substitution/Elimination Method

The substitution/elimination method is an algebraic method for solving systems of equations. It involves solving for one of the variables in one of the equations in terms of the other one, and then substituting the expression for that variable into the other equation. In this process one of the variables is eliminated, making it possible to solve for the remaining one. I recommend using this method if at least one of the coefficients of one of the variables in one of the equations is either 1 or –1; in that case one variable is easy to isolate and you don't have to complicate the problem by having to work with fractions.

Example 1

Solve the system of equations: $\begin{cases} x+2y=5 \\ 2x-y=5 \end{cases}$

Solution: Since the coefficient of x in the top equation equals 1, this method will work well. Use the first equation to solve for x:

$x + 2y = 5$

$x = -2y + 5$

I will refer to this equation as the **substitution equation**. Use this equation for x to eliminate the appearance of x in the second equation. In other words, replace every x in the second equation with the expression $-2y + 5$. After that, solve for y:

$2x - y = 5$

$2(-2y + 5) - y = 5$

$-4y + 10 - y = 5$

$-5y + 10 = 5$

$-5y = -5$

$\dfrac{\cancel{-5}y}{\cancel{-5}} = \dfrac{\cancel{-5}}{\cancel{-5}}$

$y = 1$

Now that you know the value of y, use it in the substitution equation to solve for x:

$x = -2y + 5$

$x = -2(1) + 5$

$x = 3$

The last step is to check your work. Make sure that the point (3, 1) satisfies both equations:

$\begin{cases} x+2y=5 \\ 2x-y=5 \end{cases}$

$\begin{cases} 3+2\cdot1=5 \\ 2\cdot3-1=5 \end{cases}$

Both equations are satisfied, so the system of equations $\begin{cases} x + 2y = 5 \\ 2x - y = 5 \end{cases}$ has solution (3, 1).

Example 2

Solve the system of equations $\begin{cases} 3x - 2y = 1 \\ 4x - y = -2 \end{cases}$

Solution: Because the second equation has a variable with a coefficient of –1, that is the equation that we will use to isolate y:

$4x - y = -2$

$-y = -4x - 2$

$(-1)(-y) = (-1)(-4x - 2)$

$y = 4x + 2$

Use this substitution equation to eliminate y in the first equation and solve for x:

$3x - 2y = 1$

$3x - 2(4x + 2) = 1$

$3x - 8x - 4 = 1$

$-5x - 4 = 1$

$-5x = 5$

$\dfrac{\cancel{-5}y}{\cancel{-5}} = \dfrac{\cancel{-5}}{\cancel{-5}}$

$x = -1$

Now that we know the value of x we can use the substitution equation to solve for y:

$y = 4x + 2$

$y = 4(-1) + 2$

$y = -2$

So the solution consists of the point (–1, –2). The last step is to check our solution to make sure that it satisfies both of the equations in our system:

$$\begin{cases} 3x - 2y = 1 \\ 4x - y = -2 \end{cases}$$

$$\begin{cases} 3(-1) - 2(-2) = -3 + 4 = 1 \\ 4(-1) - (-2) = -4 + 2 = -2 \end{cases}$$

Both of these equations are satisfied, so the system of equations $\begin{cases} 3x - 2y = 1 \\ 4x - y = -2 \end{cases}$ has solution (–1, –2).

The substitution/elimination method will work even if all of the coefficients of all of the variables in both of the equations are not 1 or –1. In that situation, I prefer to use the method that will be discussed in the next section. In the meantime, work out the three practice problems to make sure that you understand the methods being used. You will want to make sure you understand these problems before trying to solve the problems in the next lesson.

Lesson 7-2 Practice

Use the substitution/elimination method to solve the following systems of equations:

1. $\begin{cases} 3x + y = 7 \\ x - 2y = 0 \end{cases}$ 2. $\begin{cases} 2x + y = 5 \\ 2x - y = 0 \end{cases}$ 3. $\begin{cases} 3x - 2y = 4 \\ x - y = 2 \end{cases}$

Lesson 7-3: The Addition/Subtraction Method

This method is sometimes referred to as solving a linear system by using **linear combinations**. It should be used when all of the coefficients of the variables are real numbers other than 1 and –1. The procedure is easier to understand with an example, so I will talk you through the logic and then walk you through a couple of examples so you can see how the technique is used. Then you will have a chance to apply what you have learned.

7

SYSTEMS OF LINEAR EQUATIONS

- ▣ To begin with, you should write each equation in *standard form* and write the variables in the same order. Line up your equations one on top of the other and decide which of the variables you want to eliminate.

- ▣ Look at the coefficient of that variable in each of the two equations. Your goal is to make those coefficients the same. The best way to do that is to find the least common multiple (remember that the least common multiple is the smallest number that both coefficients divide into evenly). Figure out what you have to multiply each equation by in order to get those coefficients to be the same.

- ▣ Once you multiply each equation by it's particular constant, you are ready to add or subtract the two equations, depending on the signs. If the two coefficients are the same sign, subtract one equation from the other; if the two coefficients are opposite in sign, add the sides of the two equations together. After you do that you will be left with only one variable.

- ▣ Solve for that variable and then use that value in either one of the original equations to solve for the value of the other variable. Let's see the method in action.

Example 1

Solve the system of equations: $\begin{cases} 4x + 3y = 18 \\ 2x + 5y = 16 \end{cases}$

Solution: Each equation is already written in standard form. Notice that the coefficients of x are 4 and 2; the coefficients of y are 3 and 5. The least common multiple of 4 and 2 is 4; the least common multiple of 3 and 5 is 15. If I wanted to get rid of x I would want the coefficient of x to be 4 in both equations; if I wanted to eliminate y, I would want the coefficient of y to be 15 in both equations. I will have to do less work if I eliminate x, so that is the variable I will go after. The way to have both equations have the same coefficient for x is to multiply the second equation by 2 (and leave the first equation alone, because the coefficient of x is already 4 in the first equation). When we multiply the second equation by 2, we will still have an equivalent equation, and hence an equivalent system of equations:

$$\begin{cases} 4x + 3y = 18 \\ 4x + 10y = 32 \end{cases}$$

Notice that when I multiplied the second equation by 2, I had to multiply each term in the equation by 2. Remember that mathematics is an egalitarian system: whatever you do to one side of an equation you *must* also do to the other side in order to keep the scales balanced. Now the system that we are solving is:

$$\begin{cases} 4x + 3y = 18 \\ 4x + 10y = 32 \end{cases}$$

Because the two coefficients in front of the variable x are the same, we can subtract the bottom equation from the top equation to get:

$$\begin{array}{r} 4x + 3y = 18 \\ - \ 4x + 10y = 32 \\ \hline 0x - 7y = -14 \end{array}$$

Be sure that when you subtract the bottom equation that you subtract each and every term. You are using the distributive property: on the left-hand side of the equation you are evaluating $4x + 3y - (4x + 10y)$ and on the right-hand side of the equation you are evaluating $18 - 32$. Now we can solve directly for y by dividing both sides of the equation by -7:

$$-7y = -14$$

$$\frac{\cancel{-7}y}{\cancel{-7}} = \frac{-14}{-7}$$

$$y = 2$$

Now that we know the value of y, we can use it to substitute into either of the original equations and solve for x. I will use the first equation to find x:

$4x + 3y = 18$

$4x + 3 \cdot 2 = 18$

$4x + 6 = 18$

$4x = 12$

$\dfrac{\cancel{4}x}{\cancel{4}} = \dfrac{12}{4}$

$x = 3$

So the solution to the system of equations is the point (3, 2). The last thing to do is to check the solution in the second equation. Plug in the value of y into the second equation:

$4x + 10y = 32$

$4 \cdot 3 + 10 \cdot 2 = 32$

So the point (3, 2) satisfies both equations and is therefore the solution.

Example 2

Solve the system of equations: $\begin{cases} 2x - 3y = 0 \\ 3x - 2y = 5 \end{cases}$

Solution: Because none of the coefficients are equal to 1 or −1, the method to use would be the addition/subtraction method. The numbers we pick to multiply the first and second equation by depend on which variable you want to eliminate. In order to eliminate x from the system you would need to multiply the first equation by 3 and the second equation by 2. If you wanted to eliminate y from the system you would need to multiply the first equation by 2 and the second equation by 3. In the last example I eliminated x, so in this example I will eliminate y. First I will transform both equations:

$$\begin{cases} 2x - 3y = 0 \\ 3x - 2y = 5 \end{cases}$$

Multiply the top equation by 2:

Multiply the bottom equation by 3:

$$\begin{cases} 4x - 6y = 0 \\ 9x - 6y = 15 \end{cases}$$

Now that the coefficients in front of the variable y are the same, I will carefully subtract the bottom equation from the top of the equation:

$$\begin{array}{r} 4x - 6y = 0 \\ -9x - 6y = 15 \\ \hline -5x + 0y = -15 \end{array}$$

Finally, use the resulting equation to solve for x:

$$-5x = -15$$

$$\frac{\cancel{-5}x}{\cancel{-5}} = \frac{-15}{-5}$$

$$x = 3$$

Now that I have a value for x, I can use either of the original equations to solve for y. I will plug the value of x into the original second equation and solve for y:

$$3x - 2y = 5$$

$$3 \cdot 3 - 2y = 5$$

$$9 - 2y = 5$$

$$-2y = -4$$

$$y = 2$$

So the solution to the system of equations is the point (3, 2). The last step is to check our answer by substituting our point into the first equation:

$$2x - 3y = 0$$

$$2 \cdot 3 - 2 = 0$$

Because the point (3, 2) satisfies both equations, it is the solution to the system of equations.

Work out the following problems to be sure that you understand the technique used and can successfully apply it. Be sure to check your answers in the appendix before moving on to the next chapter.

7

SYSTEMS OF LINEAR EQUATIONS

Lesson 7-3 Practice

Use the addition/subtraction method to solve the following systems of equations:

1. $\begin{cases} 3x + 4y = 1 \\ 2x - 3y = -5 \end{cases}$

2. $\begin{cases} 2x + 3y = 3 \\ 3x - 2y = 11 \end{cases}$

3. $\begin{cases} 3x + 5y = -1 \\ 4x + 3y = -5 \end{cases}$

Answer Key
Lesson 7-1

1. Graph of the system of equations $\begin{cases} x - 2y = 0 \\ 2x + y = 10 \end{cases}$

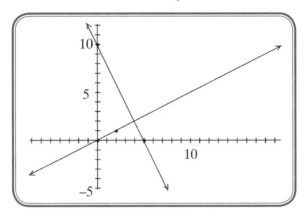

2. Graph of the system of equations $\begin{cases} 2x + 3y = 13 \\ x - y = -1 \end{cases}$

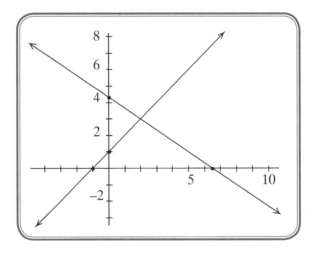

3. Graph of the system of equations $\begin{cases} x - y = -3 \\ x + y = 1 \end{cases}$

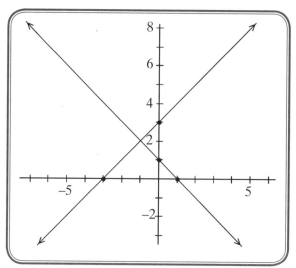

Lesson 7-2

1. $(2, 1)$

2. $\left(\dfrac{5}{4}, \dfrac{5}{2}\right)$

3. $(0, -2)$

Lesson 7-3

1. $(-1, 1)$

2. $(3, -1)$

3. $(-2, 1)$

8

Monomials and Polynomials

Your experiences with functions have exposed you to some mathematical expressions that are very important in algebra. For example, the function $f(x) = x^2$ is the function that takes whatever is in parentheses and squares it. Functions will involve numbers and variables that are combined by using the four basic operations. Algebra is all about building on the foundation of numbers, and numbers can be added, subtracted, multiplied and divided. Because functions are directly related to numbers, it should not surprise you that *functions* can also be added, subtracted, multiplied and divided. In this chapter we will begin to explore the algebra of *functions*.

Lesson 8-1: Terminology

The expression x^2 is an example of a monomial. A **monomial** is an expression that only involves a product of numbers and variables whose powers are positive integers. A monomial in one variable is an expression of the form $k \cdot x^n$ where k is a real number and n is a positive integer. The

degree of a monomial in one variable is the power that the variable is raised. The degree of the monomial $k \cdot x^n$ is n. A constant can be viewed as a monomial of the form $k \cdot x^0$ (since any non-zero number raised to the 0th power is 1), so a constant is a monomial of degree 0.

A monomial with two variables is an expression of the form $k \cdot x^n \cdot y^m$. The number k is called the **coefficient** of the monomial. The **degree** of a monomial in more than one variable is the sum of the powers of all of the variables that appear. A monomial of the form $k \cdot x^n \cdot y^m$ has degree $n + m$. The monomials $3x^4$, $5xy$, $7x^2 y^4$ have degree 4, 2, and 6, respectively.

A **polynomial** is the sum of two or more monomials. The sign of k determines whether the monomial is added to or subtracted from the other monomials in the chain. The monomials that make up a polynomial are usually written in descending order based on the degree of the monomial. In other words, you start writing a polynomial by writing the monomial with the highest degree first, then the second

Monomial or Polynomial	Leading Coefficient	Degree	Common Name	Type
4	4	0	Constant	Monomial
$2x$	2	1	Linear	Monomial
$3x - 9$	3	1	Linear	Binomial
x^2	1	2	Quadratic	Monomial
$x^2 + x$	1	2	Quadratic	Binomial
$x^2 + 3x - 9$	1	2	Quadratic	Trinomial
$-2x^3$	-2	3	Cubic	Monomial
$-5x^3 + 3x + 1$	-5	3	Cubic	Trinomial

highest degree, and so on. This is considered the standard form of a polynomial. When a polynomial is written in standard form, the coefficient of the first term is called the **leading coefficient**. The polynomials $3x^2 + 2$, $4xy + y + 1$ and $x^3 + x^2 + 1$ are written in standard form, and their leading coefficients are 3, 4, and 1, respectively.

Some polynomials have special names. A polynomial that is made up of two monomials is called a **binomial**. A polynomial that is made up of three monomials is called a **trinomial**. The table on page 162 gives examples of some of the more common types of monomials and polynomials that you will encounter in algebra.

Lesson 8-2: Adding and Subtracting Monomials

The only time that you can effectively combine (by addition or subtraction) two monomials is when both monomials involve the exact same variables raised to the exact same powers. The only possible variation in the monomials is in their coefficients. For example, you can add the monomials $2x^2y^3$ and $5x^2y^3$ together to get $7x^2y^3$, because x is squared and y is cubed in both monomials. If you try to add the monomials $2x^2y^3$ and $5x^2y^2$ together you will get $2x^2y^3 + 5x^2y^2$. There is no further simplification that you can do at this point because y is cubed in the first monomial and y is squared in the second monomial. I will work out a few examples to show you some monomials that can and cannot be combined.

Example 1

Combine the following monomials, if possible:

a. $5xy + 9xy$

b. $3x^2 + x^2$

c. $12xyz - 16xyz$

d. $12xy^2 - 3xy$

Solution:

a. The only difference in the two monomials is the coefficient, so it is possible to add them together: $5xy + 9xy = 14xy$

b. The powers of x match in each of the monomials, so it is possible to add them together: $3x^2 + x^2 = 4x^2$

c. The powers of x, y, and z are the same in both monomials, so it is possible to write: $12xyz - 16xyz = -4xyz$

d. No further combination is possible because the powers of y are different in the two monomials: $12xy^2 - 3xy = 12xy^2 - 3xy$

Lesson 8-2 Practice

Combine the following monomials:

1. $4x - 6x$

2. $13xy^2 + 2xy$

3. $3xz + 5xz$

4. $6xy - 9xy$

Lesson 8-3: Multiplying and Dividing Monomials

In Chapter 2 I promised you that the rules for exponents would play an important role in Chapter 8. If these rules don't sound familiar, it may be worth your while to go back to Chapter 2 for a quick review. I will remind you of the product, quotient, and power rules for exponents here, so you can judge whether it is best to go on or to review:

▣ **Product rule:** When you multiply two exponential expressions that have the same base, you add the exponents:

$$a^m \times a^n = a^{m+n}$$

▣ **Quotient rule:** When you divide one exponential expression into another exponential expression with the same base, you subtract the exponents:

$$\frac{a^m}{a^n} = a^{m-n}$$

▣ **Power rule:** When you raise an exponential expression to a power, you multiply the exponents:

$$\left(a^m\right)^n = a^{m\times n}$$

Multiplying monomials is more relaxed than adding or subtracting them. In order to add or subtract monomials we had to make sure that the variables in each monomial were exactly the same, down to the power. The nice thing about multiplying monomials is that it doesn't matter which variables are involved in the monomials or what their powers are. All that matters is that the only operation involved is multiplication. The strategy for multiplying monomials is as follows:

▣ First, multiply the coefficients as a group.

▣ Then, focus on each variable involved in the product.

Use the rules for how to multiply exponential expressions and you can't go wrong. The following example should help illustrate the process.

Example 1

Simplify the following expressions:

a. $\left(x^2\right)\left(2x^4\right)$

b. $\left(4x^4\right)\left(3x^6\right)$

c. $\left(x^3y^4\right)\left(-xy^2\right)$

d. $\left(-4x^4\right)\left(-6xy\right)\left(3y^2\right)$

Solution:

a. $\left(x^2\right)\left(2x^4\right) = \left(1\cdot x^2\right)\left(2x^4\right) = \left(1\cdot 2\right)\left(x^2x^4\right) = 2x^6$

b. $\left(4x^4\right)\left(3x^6\right) = \left(4\cdot 3\right)\left(x^4x^6\right) = 12x^{10}$

c. $\left(x^3y^4\right)\left(-xy^2\right) = \left(1\cdot x^3y^4\right)\left(-1\cdot xy^2\right)$
$$= \left(1\cdot(-1)\right)\left(x^3x\right)\left(y^4y^2\right) = -x^4y^6$$

d. $\left(-4x^4\right)\left(-6xy\right)\left(3y^2\right)=\left(\left(-4\right)\left(-6\right)\left(3\right)\right)\left(x^4x\right)\left(yy^2\right)=72x^5y^3$

Division proceeds just like multiplication. Check for common factors of the coefficients that you can cancel, and use the rules for dividing exponential expressions to handle the variables in the expression. The next example should help you understand the process. Notice that the expressions are starting to get complicated. We can combine multiplication and division of monomials in the same expression. When that happens, I recommend that you simplify the numerator and the denominator first and *then* do the division. I have tried to break things down step by step, so you can follow all of the changes.

Example 2

Simplify the following expressions:

a. $\dfrac{2x^4}{x^2}$

b. $\dfrac{36x^6}{9x^4}$

c. $\dfrac{\left(x^3y^4\right)\left(2x^2y^3\right)}{-xy^2}$

d. $\dfrac{\left(-4x^4\right)\left(-6xy\right)}{3y^2}$

Solution:

a. $\dfrac{2x^4}{x^2}=\left(\dfrac{2}{1}\right)\left(\dfrac{x^4}{x^2}\right)=2x^2$

b. $\dfrac{36x^6}{9x^4}=\left(\dfrac{36}{9}\right)\left(\dfrac{x^6}{x^4}\right)=4x^2$

c. $\dfrac{\left(x^3y^4\right)\left(2x^2y^3\right)}{-xy^2}=\dfrac{\left(1\cdot2\right)\left(x^3x^2\right)\left(y^4y^3\right)}{-xy^2}=\dfrac{2x^5y^7}{-xy^2}=\left(\dfrac{2}{-1}\right)\left(\dfrac{x^5}{x}\right)\left(\dfrac{y^7}{y^2}\right)$

$=-2x^4y^5$

d.
$$\frac{\left(-4x^4\right)\left(-6xy\right)}{3y^2} = \frac{\left((-4)(-6)\right)\left(x^4 x\right)y}{3y^2} = \frac{24x^5 y}{3y^2} = \left(\frac{24}{3}\right)x^5\left(\frac{y}{y^2}\right)$$

$$= 8x^5 y^{-1} = \frac{8x^5}{y}$$

Try your hand at solving some of these problems. By the time you are through, you should have a solid understanding of the product and the quotient rules for exponents.

Lesson 8-3 Practice

Simplify the following expressions. Your answer should only contain positive exponents:

1. $\left(3x^2\right)\left(2x^6\right)$

2. $\left(-5x^4 y^2\right)\left(-2x^3 y\right)$

3. $\dfrac{27x^6}{12x^4}$

4. $\dfrac{-6x^3}{15x^8}$

5. $\dfrac{\left(2x^3\right)\left(-3y^3\right)}{9x^2 y}$

6. $\dfrac{\left(4xy^2\right)\left(8x^4 y^3\right)}{12x^3 y}$

Lesson 8-4: Monomials as the Base of an Exponential Expression

Now it is time to apply the power rule for exponents and write some more interesting expressions using monomials. To simplify an exponential expression that has a monomial as the base, just apply the rule:

$$\left(a \cdot b\right)^n = a^n \cdot b^n$$

If you are careful to raise each term in the base to the power n, you can't go wrong. Remember that we worked similar problems in Chapter 2.

8

MONOMIALS AND
POLYNOMIALS

MONOMIALS AND POLYNOMIALS

8

Example 1

Simplify the following expressions:

a. $\left(x^4\right)^5$

b. $\left(3x^3\right)^4$

c. $\left(xy^3\right)^6$

d. $\left(2xy^3z^5\right)^3$

Solution:

a. $\left(x^4\right)^5 = x^{20}$

b. $\left(3x^3\right)^4 = 3^4 x^{12} = 81x^{12}$

c. $\left(xy^3\right)^6 = x^6 y^{18}$

d. $\left(2xy^3z^5\right)^3 = 2^3 x^3 y^9 z^{15} = 8x^3 y^9 z^{15}$

Lesson 8-4 Practice

Simplify the following expressions:

1. $\left(3x^2\right)^3$

2. $\left(4xy^2\right)^5$

3. $\left(2x^4y^6\right)^2$

4. $\left(4x^2yz^3\right)^3$

Lesson 8-5: Adding and Subtracting Polynomials

A polynomial is a sum of monomials. Because of the commutative and associative properties of addition, adding or subtracting polynomials is equivalent to adding or subtracting monomials. Remember that you can only combine monomials that have the exact same variables raised to the exact same powers. When you add or subtract polynomials, you have to look for monomials that match up and combine them. When you are subtracting one polynomial from another polynomial you must be sure to subtract *each term* in the polynomial being subtracted. This is equivalent to distributing the negative sign throughout the polynomial that is being subtracted, as you will see in Example 1.

Example 1

Simplify the following expressions:

a. $\left(x^2+3x-2\right)+\left(3x^2-5x-6\right)$

b. $\left(3x-8y\right)+\left(5x+2y\right)$

c. $\left(x^2y-3x\right)-\left(4x^2y+3x\right)$

d. $\left(x^2+3x+2\right)-\left(2x^2-5x-6\right)$

e. $\left(3x-8y\right)-\left(x+2y\right)$

Solution:

a. Combine the terms that involve x^2, then combine the terms that involve x, and, finally, combine the constant terms together:

$$\left(x^2+3x-2\right)+\left(3x^2-5x-6\right)=4x^2-2x-8$$

b. Combine the terms that involve x, then combine the terms that involve y:

$$\left(3x-8y\right)+\left(5x+2y\right)=8x-6y$$

c. Distribute the negative in front of the polynomial being subtracted, and then combine monomials that match up:

$$\left(x^2y-3x\right)-\left(4x^2y+3x\right)=\left(x^2y-3x\right)-4x^2y-3x=-3x^2y-6x$$

d. Distribute the negative in front of the polynomial being subtracted, and then combine monomials that match up:

$$\left(x^2+3x+2\right)-\left(2x^2-5x-6\right)=\left(x^2+3x+2\right)-2x^2+5x+6$$
$$=-x^2+8x+8$$

e. Distribute the negative in front of the polynomial being subtracted, and then combine monomials that match up:

$$\left(3x-8y\right)-\left(x+2y\right)=\left(3x-8y\right)-x-2y=2x-10y$$

MONOMIALS AND POLYNOMIALS

8

Lesson 8-5 Practice

Simplify the following expressions:

1. $\left(2x^2 + 3x - 1\right) + \left(x^2 + 2x + 7\right)$

2. $3\left(x^2 + 2x + 1\right) - \left(x^2 - 2x + 1\right)$

3. $\left(x^2y + 3x + 2\right) + \left(x^2y + 6x - 1\right)$

4. $\left(3x + 2y\right) - \left(4x - 2y\right)$

Lesson 8-6: Multiplying Polynomials

In order to understand how to multiply two polynomials together, it is best to start with how to multiply a monomial and a polynomial. The key to multiplying a monomial and a polynomial together is to use the *distributive property* and the rules for multiplying numbers with the same base.

Example 1

Find the product: $3x^2\left(2x^3 + x - 8\right)$

Solution:

The $3x^2$ term gets *distributed* to each of the terms in the polynomial:

$$3x^2\left(2x^3 + x - 8\right) = \left(3x^2\right)\left(2x^3\right) + \left(3x^2\right)(x) + \left(3x^2\right)(-8)$$
$$= 6x^5 + 3x^3 - 24x^2$$

Multiplying two polynomials together uses the same principle. There is just more to keep track of.

The process of multiplying binomials together is often referred to as **FOIL** which stands for *first, outside, inside,* and *last*. This represents one way of doing the multiplication as shown in Figure 8.1.

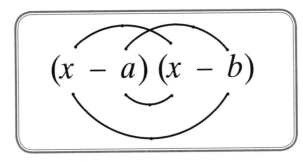

Figure 8.1: The face of FOIL.

For example, when you multiply two binomials, you are using the distributive property twice:

$$(a+b)(c+d)=a(c+d)+b(c+d)=ac+ad+bc+bd$$

It doesn't matter whether the binomials involve addition or subtraction. You distribute the sign. For the product $(a-b)(c+d)$, the negative sign in front of b tags along in front of b throughout the entire process:

$$(a-b)(c+d)=a(c+d)-b(c+d)=ac+ad-bc-bd$$

For the product $(a+b)(c-d)$, the negative sign in front of d tags along in front of d throughout the entire process:

$$(a+b)(c-d)=a(c-d)+b(c-d)=ac-ad+bc-bd$$

For the product $(a-b)(c-d)$, the negative signs in front of both b and d tag along throughout the entire process. The last step makes use of the fact that a negative times a negative equals a positive:

$$(a-b)(c-d)=a(c-d)-b(c-d)=ac-ad-bc+(-b)(-d)$$
$$=ac-ad-bc+bd$$

Example 2

Find the product: $(x + 3)(x + 2)$

Solution:

Use the distributive property twice and gather like terms:

$$(x+3)(x+2) = x(x+2)+3(x+2)$$
$$= x^2 + 2x + 3x + 6$$
$$= x^2 + 5x + 6$$

Example 3

Find the product: $(x + 3)(x - 4)$

Solution:

Use the distributive property twice and gather like terms:

$$(x+3)(2x-4) = x(2x-4)+3(2x-4)$$
$$= 2x^2 - 4x + 6x - 12$$
$$= 2x^2 + 2x - 12$$

Example 4

Find the product: $(2x-3)(x+1)$

Solution:

Use the distributive property twice and gather like terms:

$$(2x-3)(x+1) = 2x(x+1)-3(x+1)$$
$$= 2x^2 + 2x - 3x - 3$$
$$= 2x^2 - x - 3$$

Example 5

Find the product: $(x - 3)(x - 5)$

Solution:

Use the distributive property twice and gather like terms:

$$(x-3)(x-5) = x(x-5)-3(x-5)$$
$$= x^2 - 5x - 3x + 15$$
$$= x^2 - 8x + 15$$

Multiplying a binomial and a trinomial is done in the same way as multiplying two binomials. There is just another term to keep track of as you multiply things together. The key is to keep distributing until you have multiplied everything together that should be multiplied together.

Example 6

Find the product: $(x-3)(2x^2-5x+1)$

Solution:

Apply the distributive property and collect similar terms:

$$(x-3)(2x^2-5x+1)=x(2x^2-5x+1)-3(2x^2-5x+1)$$
$$=2x^3-5x^2+x-6x^2+15x-3$$
$$=2x^3-11x^2+16x-3$$

Lesson 8-6 Practice

Find the following products:

1. $4x^5(x^3+2x+1)$

2. $(x+2)(x-3)$

3. $(2x+1)(x+2)$

4. $(x-1)(x+1)$

5. $(x-1)(x^2+x+1)$

Lesson 8-7: Special Products

There are a few special products that come up frequently in algebra. If we take some time to work out a few specific examples and make some observations, it will cut our work down in later sections.

The first product we will examine is the square of a binomial. If we square the binomial $(a+b)$, we get:

$$\begin{aligned}
(a+b)^2 &= (a+b)(a+b) \\
&= a(a+b)+b(a+b) \\
&= a^2 + ab + ba + b^2 \\
&= a^2 + 2ab + b^2
\end{aligned}$$

Notice that the first and last terms in the expansion are just the squares of the first and last terms in the binomial itself. The middle term is the twice the product of the terms of the binomial expression. I've written out a few examples to help you see the pattern:

Product	Expansion
$(x+1)^2$	$x^2 + 2 \cdot x \cdot 1 + 1^2 = x^2 + 2x + 1$
$(x+2)^2$	$x^2 + 2 \cdot x \cdot 2 + 2^2 = x^2 + 4x + 4$
$(x+3)^2$	$x^2 + 2 \cdot x \cdot 3 + 3^2 = x^2 + 6x + 9$
$(x+4)^2$	$x^2 + 2 \cdot x \cdot 4 + 4^2 = x^2 + 8x + 16$
$(2x+1)^2$	$(2x)^2 + 2 \cdot 2x \cdot 1 + 1^2 = 4x^2 + 4x + 1$
$(3x+2)^2$	$(3x)^2 + 2 \cdot 3x \cdot 2 + 2^2 = 9x^2 + 12x + 4$

Because squaring binomials is common in algebra, I would recommend that you familiarize yourself with this process so that it becomes second nature. If the binomial involves subtraction instead of addition, that negative sign will only affect the middle term:

$$(a-b)^2 = (a-b)(a-b)$$
$$= a(a-b) - b(a-b)$$
$$= a^2 - ab - ba + b^2$$
$$= a^2 - 2ab + b^2$$

I'll give you a few examples so that you can see the difference the sign makes.

Product	Expansion
$(x-1)^2$	$x^2 - 2x + 1$
$(x-2)^2$	$x^2 - 4x + 4$
$(x-3)^2$	$x^2 - 6x + 9$
$(x-4)^2$	$x^2 - 8x + 16$
$(2x-1)^2$	$4x^2 - 4x + 1$
$(3x-2)^2$	$9x^2 - 12x + 4$

The second special product involves two binomials that involve the same terms but differ in one of the signs. Specifically, the product looks something like $(a+b)(a-b)$. Expanding this product gives:

$$(a+b)(a-b) = a(a-b) + b(a-b)$$
$$= a^2 - ab + ba - b^2$$
$$= a^2 - b^2$$

The expression $a^2 - b^2$ is called the **difference between two squares**. Notice that when you expand the product $(a+b)(a-b)$ the result is a binomial with no middle term. It's worth writing out a few examples, just to see the pattern clearly:

Product	Expansion
$(x+1)(x-1)$	$x^2 - 1^2 = x^2 - 1$
$(x+2)(x-2)$	$x^2 - 2^2 = x^2 - 4$
$(x+3)(x-3)$	$x^2 - 3^2 = x^2 - 9$
$(2x+1)(2x-1)$	$(2x)^2 - 1^2 = 4x^2 - 1$
$(3x+2)(3x-2)$	$(3x)^2 - 2^2 = 9x^2 - 4$

Lesson 8-7 Practice

Expand the following products:

1. $(x+10)^2$

2. $(3x-2)^2$

3. $(x-8)^2$

4. $(2x+1)^2$

5. $(x-6)(x+6)$

6. $(4x+1)(4x-1)$

Lesson 8-8: Dividing Polynomials by Monomials

The first step in dividing a polynomial by a monomial is to turn this problem into several problems that involve a monomial divided by another monomial. This is done by looking at the polynomial in the numerator as the sum of several individual monomials. Once you have turned this one problem into several smaller problems, the division can be done by using the rules of exponents.

Example 1

Simplify the ratio: $\dfrac{3x^3 + 2x^2 + x}{2x}$

Solution:

Use the rules for adding fractions to break up the polynomial:

$$\frac{3x^3 + 2x^2 + x}{2x} = \frac{3x^3}{2x} + \frac{2x^2}{2x} + \frac{x}{2x}$$

Then cancel where you can and divide using the rules for exponents:

$$\frac{3x^3}{2x} + \frac{2x^2}{2x} + \frac{x}{2x} = \frac{3}{2}x^2 + x + \frac{1}{2}$$

Example 2

Simplify the ratio: $\dfrac{12x^4 + 4x^3 + 8x^2}{2x^2}$

Solution:

Use the rules for adding fractions to break up the polynomial:

$$\frac{12x^4 + 4x^3 + 8x^2}{2x^2} = \frac{12x^4}{2x^2} + \frac{4x^3}{2x^2} + \frac{8x^2}{2x^2}$$

Then cancel where you can and divide using the rules for exponents:

$$\frac{12x^4}{2x^2} + \frac{4x^3}{2x^2} + \frac{8x^2}{2x^2} = 6x^2 + 2x + 4$$

Lesson 8-8 Practice

Simplify the following ratios:

1. $\dfrac{2x^4 + 6x^3 + x}{3x}$

2. $\dfrac{x^5 + 4x^3 + x^2}{4x^2}$

Lesson 8-9: Factoring Techniques

Now that you have learned how to use the distributive property to expand products of binomials, it's time to practice going in the opposite direction. In other words, if you have an expanded polynomial, can you write it as the product of monomials and binomials? This is similar to learning your multiplication tables first, and then learning how to draw factor trees.

The nice thing about working with numbers is that there is a systematic approach to factoring: work with prime numbers. With algebraic expressions, the concept of primes is a little less intuitive. But there is a systematic approach to factoring algebraic expressions, and I'll walk you through some of the basics.

- ▣ The first step in trying to factor a polynomial is to look at each monomial that makes up the polynomial. Look at all of the coefficients involved and see if they all share a common factor.

- ▣ Then look at the variables involved in each term. Is there a variable that is common to all of the monomials? If so, look at the power of that variable in each of the monomials. The monomial with that variable to the lowest power controls everything. You have to keep looking until each term has nothing in common with the others.

- ▣ I recommend checking your work after each step, just to be sure that you haven't made a mistake. To check your work you just multiply things out and undo all of your hard work.

Example 1

Factor: $5x + 10$

Solution:

Each coefficient in the binomial is divisible by 5, so we can factor out a 5: $5x + 10 = 5(x + 2)$. Notice that x and 2 have nothing in common, so we are done. We can use the distributive property to check our work: is $5(x + 2)$ the same thing as $5x + 10$? Yes.

Example 2

Factor: $3x^2 + 5x$

Solution:

The coefficients of each term are relatively prime, so there aren't any constants that we can factor out. Next, turn your attention to the variables: the first term involves an x, and so does the second term. The first term has x^2, the second term has x^1. The second term has the lowest power of x, so that one determines how many x's you can factor out. In this case I can factor out x^1:

$$3x^2 + 5x = x(3x + 5)$$

Now my terms are $3x$ and 5. They share no common constant factors and they share no variables, so that's all I can do. But I'll check my work just to be sure of my answer: $x(3x + 5) = 3x^2 + 5x$

Example 3

Factor: $9x^4 + 12x^2$

Solution:

The coefficients of each term are multiples of 3, so I can factor out a 3: $9x^4 + 12x^2 = 3(3x^4 + 4x^2)$. Next, look at the variables: the first term has x^4 and the second term has x^2. The smaller power of x is the one that I can factor out:

$$3(3x^4 + 4x^2) = 3x^2(3x^2 + 4)$$

The last step is to check our work by distributing:

$3x^2(3x^2 + 4) = 9x^4 + 12x^2$, so our answer is correct.

In the next chapter I will discuss factoring quadratic polynomials in much greater detail. This is just a warm-up to get you started.

Lesson 8-9 Practice

Factor the following algebraic expressions:

1. $6x - 12$

2. $6x^2 + 15x$

3. $9x^3 + 3x$

4. $12x^4 + 15x^3$

Lesson 8-10: Solving Factored Equations

Factoring polynomials is very important in algebra. In fact, it is a skill that is used in many math classes beyond algebra. You'll be surprised at how much easier life (or at least algebra) will be once you master the factoring techniques discussed in this book. One of the main reasons to factor equations is because it is much easier to solve factored equations than it is to solve un-factored equations.

Solving factored equations relies on one of the special properties of 0. Remember that 0 is the only number with the property that if $a \cdot b = 0$ then either $a = 0$, or $b = 0$. When you factor a polynomial, you are writing the polynomial as a product of its factors. If that polynomial happens to be equal to 0, then you know that the only way that a product of numbers equals 0 is if at least one of the products equals 0. So finding the values of x that make a complicated equation equal to 0 becomes a matter of setting each factor equal to 0 and solving these simpler equations.

Example 1

Solve the equation $(2x - 1)(x + 3) = 0$.

Solution: Because the product of $(2x - 1)$ and $(x + 3)$ is equal to 0, we know that either $(2x - 1) = 0$ or $(x + 3) = 0$. Solve each of these equations for x:

$$(2x - 1) = 0 \qquad\qquad (x + 3) = 0$$
$$2x = 1 \qquad\qquad\qquad x = -3$$
$$x = \frac{1}{2}$$

That means that either $x = \frac{1}{2}$, or $x = -3$.

These problems involve solving linear equations and making use of the property of 0. Having thoroughly learned both of these ideas, these new problems are no problem!

Example 2

Solve the equation: $(3x+2)(x-5)(x+1)=0$

Solution: It doesn't matter whether there are two factors, three factors, or a million factors. The process remains the same: set each factor equal to 0 and solve for x:

$(3x+2)=0$ $\qquad$ $(x-5)=0$ $\qquad$ $(x+1)=0$

$\qquad 3x=-2$ $\qquad\qquad x=5$ $\qquad\qquad x=-1$

$$x=-\frac{2}{3}$$

So either $x=-\frac{2}{3}$, $x=5$, or $x=-1$.

Lesson 8-10 Practice

Solve the following equations:

1. $(3x+1)(x-2)=0$
2. $(2x-4)(x+1)=0$
3. $(x+1)(x-2)(2x-1)=0$
4. $(x+4)(x-3)(x+5)=0$

Answer Key

Lesson 8-2

1. $-2x$

2. $13xy^2 + 2xy$

3. $8xz$

4. $-3xy$

Lesson 8-3

1. $6x^8$

2. $10x^7y^3$

3. $\dfrac{9x^2}{4}$

4. $\dfrac{2}{5x^5}$

5. $\dfrac{-2xy^2}{3}$

6. $\dfrac{8x^2y^4}{3}$

Lesson 8-4

1. $27x^6$

2. $1{,}024x^5y^{10}$

3. $4x^8y^{12}$

4. $64x^6y^3z^9$

Lesson 8-5

1. $3x^2 + 5x + 6$

2. $2x^2 + 8x + 2$

3. $2x^2 + 9x + 1$

4. $-x + 4y$

Lesson 8-6

1. $4x^8 + 8x^6 + 4x^5$

2. $x^2 - x - 6$

3. $2x^2 + 5x + 2$

4. $x^2 - 1$

5. $x^3 - 1$

Lesson 8-7

1. $x^2 + 20x + 100$

2. $9x^2 - 12x + 4$

3. $x^2 - 16x + 64$

4. $4x^2 + 4x + 1$

5. $x^2 - 36$

6. $16x^2 - 1$

Lesson 8-8

1. $\dfrac{2x^3}{3} + 2x^2 + \dfrac{1}{3}$

2. $\dfrac{x^3}{4} + x + \dfrac{1}{4}$

Lesson 8-9

1. $6(x - 2)$

2. $3x(2x + 5)$

3. $3x(3x^2 + 1)$

4. $3x^3(4x + 5)$

Lesson 8-10

1. $x = -\dfrac{1}{3}$ or $x = 2$

2. $x = 2$ or $x = -1$

3. $x = -1$ or $x = 2$ or $x = \dfrac{1}{2}$

4. $x = -4$ or $x = 3$ or $x = -5$

9

Quadratic Equations

Quadratic expressions are algebraic expressions of the form $ax^2 + bx + c$, and **quadratic equations** are equations of the form $ax^2 + bx + c = 0$. With both quadratic expressions and quadratic equations, the value of the leading coefficient, a, cannot be 0. The other coefficients b and c can be 0, but the leading coefficient cannot. Because quadratic expressions involve squaring numbers, solving quadratic equations sometimes involves finding the square root of a number. So before we delve into the quadratics we must take a detour and study square roots.

Lesson 9-1: Square Roots

Squaring a number involves multiplying a number by itself. For example, the square of 4 is $4^2 = 16$; the square of -3 is $(-3)^2 = 9$. Notice that squaring a positive number gives a positive number, and squaring a negative number also results in a positive number. The square of 0 is $0^2 = 0$. So if you square any real number (whether it is positive, negative or 0) you will get a number that is either greater than 0 or equal to 0, but never less than 0. In other words, if a is any real number, then $a^2 \geq 0$. There is no way that the square of a real number can be negative.

Finding the square root of a number involves reversing the process: if $b^2 = a$, then b is called the **square root** of a. For example, $4^2 = 16$ so 4 is the square root of 16. Notice that $(-4)^2 = 16$ so –4 is also the square root of 16. In fact, all positive real numbers have two square roots: a positive square root, sometimes called the **principal square root**, and a negative square root. The two square roots have the same magnitude but are opposite in sign. Square roots are written using a "square root" symbol: $\sqrt{}$. This symbol is called a **radical**. The number underneath the radical is called the **radicand**. To write things more compactly, the square roots of a number are sometimes written together as $\pm\sqrt{}$. We could write the two square roots of 16 as ± 4. Be careful with this notation and the terminology. If we write $\sqrt{4}$ we are limiting ourselves to the positive square root of 4, whereas, in general, 4 has two square roots: $\pm\sqrt{4}$. The radical by itself refers to the principal (or positive) square root.

Simplifying a square root involves factoring the radicand and pulling out the "perfect squares." **Perfect squares** are numbers whose square roots are integers. For example, since $4^2 = 16$ we could write $\sqrt{16} = 4$. Because the square root of 16 is 4 (an integer), we say that 16 is a perfect square. It is easy to list the perfect squares: just start squaring the natural numbers. The first ten perfect squares are 1, 4, 9, 16, 25, 36, 49, 64, 81, 100. To simplify a radical you need to factor the radicand and look for perfect squares. Then make use of the rule that if a is positive, then $\sqrt{a^2} = a$.

Square roots follow the same rules as exponents. In particular, if a and b are positive numbers then $\sqrt{a \cdot b} = \sqrt{a} \cdot \sqrt{b}$. We can use this relationship to simplify square roots by pulling out the perfect squares: $\sqrt{a^2 \cdot b} = \sqrt{a^2} \cdot \sqrt{b} = a\sqrt{b}$.

Example 1

Simplify: $\sqrt{64}$

Solution: Because 64 is a perfect square (you can see it in the list: $8^2 = 64$) we can write $\sqrt{64} = 8$.

Example 2

Simplify: $\sqrt{50}$

Solution: See if 50 has any perfect squares as factors. Since $50 = 25 \cdot 2$ we can write $\sqrt{50} = \sqrt{25 \cdot 2} = \sqrt{25} \cdot \sqrt{2} = 5\sqrt{2}$.

Example 3

Simplify: $\sqrt{120}$

Solution: Factor 120 into products of perfect squares: $120 = 4 \cdot 30$ so $\sqrt{120} = \sqrt{4 \cdot 30} = \sqrt{4} \cdot \sqrt{30} = 2\sqrt{30}$.

The easiest way to approach these problems is to look at the list of perfect squares and check to see if any of them divide the radicand evenly. Like most ideas in math, the more you work with perfect squares, the more you will recognize them without having to rely on the list. If there is a perfect square that evenly divides into the radicand, factor the radicand into a product of the perfect square and its corresponding factor:

Radicand = Perfect Square × Other Factor

Now do the same thing with the other factor: see if any other perfect squares divide that other factor. If they do, keep factoring. If not, pull all of the perfect squares that you found out of the radical.

It turns out that the only way for the square root of a number to be an integer is if it is a perfect square. The square roots of any other kind of number will be an irrational number.

We can also evaluate expressions that involve square roots. For example, consider the expression $\sqrt{b^2 - 4 \cdot a \cdot c}$. We can substitute

9

QUADRATIC EQUATIONS

various values for a, b and c and evaluate the expression. In this expression, the parentheses are invisible, or implied; it should be thought of as $\sqrt{(b^2 - 4 \cdot a \cdot c)}$. This expression is so useful that I thought we should take some time now to play with it.

Example 4

Evaluate the expression $\sqrt{b^2 - 4 \cdot a \cdot c}$ for $a = 2, b = 3$, and $c = -2$.

Solution: Substitute the values of the variables and work it out. Remember that you must do the addition/subtraction first, and then take the square root:

$$\sqrt{b^2 - 4 \cdot a \cdot c} = \sqrt{3^2 - 4 \cdot 2 \cdot (-2)}$$
$$= \sqrt{9 + 16}$$
$$= \sqrt{25}$$
$$= 5$$

Lesson 9-1 Practice

Answer the following questions.

1. Simplify the following radicals:
   a. $\sqrt{63}$
   b. $\sqrt{80}$
   c. $\sqrt{108}$

2. Evaluate and simplify the expression $\sqrt{b^2 - 4 \cdot a \cdot c}$ for a = –3, b = 11, and c = 12.

Lesson 9-2: The Difference Between Two Squares

In general, a quadratic expression looks something like $ax^2 + bx + c$, but we'll start out by assuming that $b = 0$. In other words, we will first turn our attention to a special group of quadratic expressions that have no middle term, or no variable with degree 1: $ax^2 + c$. Now, if both a and c are the same sign (either both are positive or both are negative) then there is nothing you can do with this expression. You can't factor it, and if you set it equal to 0 you can't solve it. To understand why, let's take a specific example: $x^2 + 1$. Set this expression equal to 0 and try to solve it:

$x^2 + 1 = 0$

$x^2 = -1$

Hmmmm......x is a real number whose square is negative. We just got through studying squares of numbers and observed that the square of any real number is greater than or equal to 0; –1 is less than 0, so that's a problem. The only way around this problem is if we declare that there are no real solutions to the equation $x^2 + 1 = 0$.

Quadratic equations of the form $ax^2 + c = 0$ will only have solutions if the coefficients a and c have opposite signs. And if the coefficients a and c have opposite signs, then the quadratic expression $ax^2 + c$ will look like the difference between two squares that we discussed in Chapter 8.

Recall that when we multiplied two factors of the form $(a + b)(a - b)$ we ended up with a result that we called the difference between two squares:

$(a + b)(a - b) = a^2 - b^2$

So we can work backwards and factor expressions that are the difference between two squares:

$a^2 - b^2 = (a + b)(a - b)$

Why would we want to factor the difference between two squares after we went through all of the trouble to multiply it out and make our observations? Because algebraic equations can be solved easily if the equations are written as a product of terms that multiply to give 0. And writing expressions as products involves factoring. So we will learn how to factor, starting with expressions that involve the difference between two squares.

Example 1

Factor the expression: $x^2 - 4$

Solution: $x^2 - 4 = (x + 2)(x - 2)$

Example 2

Factor the expression: $x^2 - 25$

Solution: $x^2 - 25 = (x+5)(x-5)$

Example 3

Factor the expression: $x^2 - 3$

Solution: $x^2 - 3 = \left(x + \sqrt{3}\right)\left(x - \sqrt{3}\right)$

Notice that the difference between two squares doesn't have to involve perfect squares. Any positive number can be thought of as the square of a number, since $a = \sqrt{a^2}$. All you have to do is take the square root of each of the terms and follow the pattern.

Example 4

Factor the expression: $4x^2 - 5$

Solution: $4x^2 - 5 = \left(2x + \sqrt{5}\right)\left(2x - \sqrt{5}\right)$

Lesson 9-2a Practice

Factor the following expressions:

1. $x^2 - 36$

2. $x^2 - 15$

3. $9x^2 - 4$

Now that we have mastered factoring the difference between two squares, it is time to use that technique to solve some quadratic equations. We'll make use of the fact that in order for a product to equal 0, at least one of the factors must be equal to 0. By factoring our quadratic expression and using this important property of 0 we can solve many quadratic equations.

Example 5

Solve the equation: $x^2 - 16 = 0$

Solution: Factor the expression, set each factor equal to 0 and solve it:

$$x^2 - 16 = 0$$
$$(x+4)(x-4) = 0$$

$$(x+4) = 0 \qquad\qquad (x-4) = 0$$
$$x = -4 \qquad\qquad\qquad x = 4$$

There are two solutions to this equation: $x = \pm 4$

Example 6

Solve the equation: $x^2 = 12$

Solution: The key to solving equations is to set a product equal to 0. There is not a 0 to be found in this equation. That's okay, though. Look at what happens when we subtract 12 from both sides of the equation:

$x^2 = 12$

$x^2 - 12 = 0$

Now we have an equation in a familiar form (the difference between two squares) so we can solve it:

$$x^2 - 12 = 0$$
$$\left(x+\sqrt{12}\right)\left(x-\sqrt{12}\right) = 0$$

$$\left(x+\sqrt{12}\right) = 0 \qquad\qquad \left(x-\sqrt{12}\right) = 0$$
$$x = -\sqrt{12} \qquad\qquad\qquad x = \sqrt{12}$$

There are two solutions to this equation: $x = \pm\sqrt{c}$

Notice that the solutions to the equation $x^2 = 12$ are $x = \pm\sqrt{c}$. We can generalize this observation: the equation $x^2 = c$ has solutions

$x = \pm\sqrt{c}$. This generalization is the basis for the process "take the square root of both sides and throw in the $\pm$ sign." We can apply this general rule to the problem in Example 6: $x^2 = 12$. On the left side of the equation, we have the perfect square x^2, so its square root is just x. On the right side of the equation, we have the number 12, and its square root is just $\sqrt{12}$. Toss in the $\pm$ sign and we're done: $x = \pm\sqrt{c}$ This simplified process cuts to the chase; it gives steps for how to obtain the solution quickly while masking the reasons why the process works, but we can use this generalization to help us solve more complicated equations.

Example 7

Solve the equation: $(x+1)^2 = 9$

Solution: Using our earlier observation, $(x+1)^2$ is a perfect square with square root $(x+1)$. On the right, we have another perfect square: 9. Its square root is 3. Toss in the $\pm$ and we're almost done: $(x+1) = \pm 3$. We're not quite finished because the goal was not to solve for $(x+1)$ but rather to solve for x. The equation $(x+1) = \pm 3$ is really two equations: $x+1 = 3$ and $x+1 = -3$. We need to solve each equation separately:

$$x+1 = 3 \qquad\qquad x+1 = -3$$
$$x = -1+3 \qquad\qquad x = -1-3$$
$$x = 2 \qquad\qquad x = -4$$

So we have two solutions to our equation: $x = 2$ or $x = -4$. We can check to make sure that both solutions work: if $x = 2$ then $(2+1)^2 = 3^2 = 9$ and if $x = -4$ then $(-4+1)^2 = (-3)^2 = 9$. So both solutions work.

In general, we can apply this process to solve equations of the form $(x-a)^2 = c$:

$$(x-a)^2 = c$$
$$(x-a) = \pm\sqrt{c}$$

$$x = a \pm \sqrt{c}$$

Lesson 9-2b Practice

Solve the following equations:

1. $x^2 = 81$
2. $x^2 = 27$
3. $(x+2)^2 = 4$
4. $(x-2)^2 = 10$

Lesson 9-3: Factoring Quadratic Expressions of the Form $x^2 + bx + c$

In order to factor a quadratic expression of the form $x^2 + bx + c$ where b and c are integers, we need to find two linear factors, say $x + r$ and $x + s$, whose product gives the expression $x^2 + bx + c$. We can expand the product $(x+r)(x+s)$ and compare that result to our expression $x^2 + bx + c$ to determine the values of r and s:

$$(x+r)(x+s) = x^2 + (r+s)x + rs$$

We want the product $(x+r)(x+s)$ to equal $x^2 + bx + c$, which means that $r + s = b$ and $r \cdot s = c$.

By trial and error, we need to find two integers, r and s, whose sum is b and whose product is c. There are some clues to help us home in on the values of r and s. It is best to think about this problem in two parts: finding the *signs* of r and s and finding the *magnitude* of r and s. We can break things up into two cases, depending on the sign of c.

◻ If c is positive, the r and s must have the same sign: either both will be positive or both will be negative. Their actual sign will depend on the sign of b:

1. If b is positive, then r and s are positive.
2. If b is negative, then r and s are negative.

So, if c is positive, r and s have the same sign that b has, and you will be looking for two numbers whose product is c and whose sum is b.

□ If c is negative, then r and s must have opposite signs.

1. The larger of r and s will have the same sign as b,
2. The smaller of r and s will the opposite sign of b.

So if c is negative the magnitude of r and s will be two positive numbers whose product is c and whose *difference* is b. I'll summarize this in a table to help you keep it straight.

Sign of c	Sign of b	Sign of r	Sign of s
+	+	+	+
+	−	−	−
−	+	Larger of $\|r\|$ and $\|s\|$ is positive, the other is negative	
−	−	Larger of $\|r\|$ and $\|s\|$ is negative, the other is positive	

You are looking for values r and s whose sum (or difference, if c is negative) is b and whose product is c. That may sound difficult at first, but after you work through some problems you'll start to get the hang of it. Of course, if c is a prime number, there aren't many integers whose product is c: just c and 1. So, in that case, the problems are fairly straightforward. Things can get a bit tricky when c is a composite number.

Example 1

Factor the expression: $x^2 + 6x + 5$

Solution: We are looking for r and s that satisfy the equation

$$x^2 + 6x + 5 = x^2 + (r+s)x + rs .$$

Because c is positive (it is +5), both r and s have to have the same sign. Because b is positive (it is +6), both r and s have to be positive. We are looking for two positive integers whose product is 5 and whose sum is 6. Because 5 is prime, our only options for r and s are 1 and 5; fortunately, they work: $1 \cdot 5 = 5$ and $1 + 5 = 6$. It doesn't matter which value you pick for r and which value you pick for s. The important thing is that their product is 5 and their sum is 6. The linear factors are $(x+1)(x+5)$. Check your answer by multiplying these two binomials together to see if I'm correct: $(x+1)(x+5)=x^2+6x+5$.

Example 2

Factor the expression: x^2-8x+7

Solution: Because c is positive (it is +7), both r and s have to have the same sign. Because b is negative (it is –8), both r and s have to be negative. We are looking for two negative integers whose product is 7 and whose sum is –8. Since 7 is prime, our only options are –1 and –7, and they work: $(-1) \cdot (-7) = 7$ and $-1 + (-7) = -8$. So the linear factors are $(x-1)(x-7)$. Multiply these two binomials together to see if I'm correct: $(x-1)(x-7)=x^2-8x+7$.

Example 3

Factor the expression: x^2-6x-7

Solution: Because c is negative, we know that r and s must differ in sign. So $r+s=-6$ and $r \cdot s=-7$; $r=-7$ and $s=1$ works: $x^2-6x-7=(x-7)(x+1)$

Example 4

Factor the expression: x^2-5x+6

Solution: Because c is positive, both r and s have the same sign. Because b is negative, both r and s are negative. In this case, however, c is a composite number: 6 can be factored as $(-1) \cdot (-6)$ or as $(-2) \cdot (-3)$. You need two factors of 6 that add up to –5; –2 and –3 are the values of r and s:

$$x^2-5x+6=(x-2)(x-3)$$

The more familiar you are with factoring whole numbers, the easier it will be to factor quadratic expressions. Sometimes you have to use a trial and error method: try all possible combinations and see which ones work. You can always check your answers by multiplying everything out.

This factoring technique works well when the linear factors have integer coefficients. If that *always* happened, then this chapter on factoring would be short. There is a quick way to determine whether the linear factors of a quadratic expression have integer coefficients, and it involves evaluating the expression $b^2 - 4 \cdot a \cdot c$, where a, b, and c are the coefficients of the quadratic expression $ax^2 + bx + c$. If the expression $b^2 - 4 \cdot a \cdot c$ is a perfect square, then the factors will have integer coefficients. For example, we saw that the quadratic expression $x^2 - 5x + 6$ could be factored into two binomials with integer coefficients. Let's evaluate $b^2 - 4 \cdot a \cdot c$ for this expression. In this case $a = 1, b = -5$, and $c = 6$, so $b^2 - 4 \cdot a \cdot c = 25 - 4 \cdot 1 \cdot 6 = 25 - 24 = 1$.

Notice that 1 is a perfect square.

> The expression $b^2 - 4 \cdot a \cdot c$ is so useful that it has a name: it is called the **discriminant**.

The linear factors of a quadratic polynomial will have integer coefficients only if the discriminant is a perfect square.

Example 5

Evaluate the discriminant of the quadratic polynomial $x^2 + 2x - 2$ and determine whether its linear factors have integer coefficients.

Solution: Evaluate the discriminant, $b^2 - 4 \cdot a \cdot c$, for the polynomial $x^2 + 2x - 2$. In the polynomial $x^2 + 2x - 2$, $a = 1, b = 2$, and $c = -2$, and the discriminant is:

$$b^2 - 4 \cdot a \cdot c = 2^2 - 4 \cdot 1 \cdot (-2) = 4 + 8 = 12.$$

Because 12 is not a perfect square, we know that the linear factors will not have integer coefficients.

Lesson 9-3 Practice

1. Factor the following quadratic expressions:

    a. $x^2 - x - 2$ c. $x^2 + 5x + 4$

    b. $x^2 + 5x - 6$ d. $x^2 - 6x + 8$

2. Evaluate the discriminant for the following polynomials and determine whether their linear factors have integer coefficients or not.

    a. $x^2 - 3x - 4$ b. $x^2 + 3x - 2$

Lesson 9-4: Factoring Quadratic Expressions in General

We will now turn our attention to factoring quadratic expressions whose leading coefficient is not 1. To factor an expression of the form $ax^2 + bx + c$ you must find the factors of the leading coefficient, a, (call them m and n) and the factors of c (call them s and t) so that the product $(mx + s)(nx + t)$ is $ax^2 + bx + c$. Expanding the product $(mx + s)(nx + t)$ and comparing it to $ax^2 + bx + c$, we see that

$$(mx + s)(nx + t) = (mn)x^2 + (mt + ns)x + (st).$$

From this we can see that the sum of the mixed products $mt + ns$ must equal b. The first couple of times that you factor these types of expressions, you will probably do some sort of trial and error approach: try possible combinations of these mixed products until you find the winner.

Example 1

Factor: $2x^2 + 11x + 5$

Solution: Mix and match the possible factors of 2 (1 and 2) and the possible factors of 5 (1 and 5) so that the sum is 11; you can mix these factors up two ways:

$1 \cdot 1 + 2 \cdot 5 = 1 + 10 = 11$

$1 \cdot 5 + 2 \cdot 1 = 5 + 2 = 7$

The first combination gives 11 and the second combination gives 7. The middle term in our quadratic expression is 11, so we want the first combination. Another way to look at this problem is to mix and match the factors and form two binomial products: $(1x+1)(2x+5)$ and $(2x+1)(1x+5)$. Expand each product and choose the one that matches the quadratic expression you set out to factor:

$(1x+1)(2x+5)=2x^2+7x+5$, which is not what we want.

$(2x+1)(1x+5)=2x^2+11x+5$, which is what we want.

Either way, we can factor the quadratic expression $2x^2+11x+5$ as $(2x+1)(1x+5)$.

Example 2

Factor: $3x^2-4x-7$

Solution: First, notice that the sign of c (the constant term) is negative and the sign of b (the middle term) is also negative. That means that our linear factors will involve different signs. We will have to mix up the factors that get multiplied together as well as the signs associated with each product. We will need to combine the factors of 3 (1 and 3) and the factors of 7 (1 and 7) so that their difference is -4. Use a trial and error method and try all possible combinations and stop when you find the one that gives you $3x^2-4x-7$:

$(3x+1)(x-7)=3x^2-20x-7$	Not equal to $3x^2-4x-7$
$(3x-1)(x+7)=3x^2+20x-7$	Not equal to $3x^2-4x-7$
$(3x+7)(x-1)=3x^2+4x-7$	Not equal to $3x^2-4x-7$
$(3x-7)(x+1)=3x^2-4x-7$	Is equal to $3x^2-4x-7$

So the correct factorization of $3x^2-4x-7$ is $(3x-7)(x+1)$.

The more practice you have expanding products of binomials the quicker you will be able to see the right combination that works. Don't feel like you have to try every possible product before deciding on the right factors. Once you find two binomials whose product is what you were given, write your answer down and move on to the next problem. I am providing all possible combinations solely for your benefit.

The first two examples were special because both a and c were prime; there was only one way to factor a and c, so our choices were limited. If a or c are composite numbers, then you have to mix and match *all* combinations of *all* possible ways to factor a and c. These problems can get a bit long, but the good news is that you will get lots of practice multiplying binomials together.

Example 3

Factor: $3x^2 + 2x - 8$

Solution: Notice that c is negative and b is positive, so the constant terms in each factor must have different signs. Also, a is prime but c is composite. Mix and match the factors of 3 (1 and 3) and the two ways to factor 8 ($1 \cdot 8$ and $2 \cdot 4$) together with the different arrangements of the signs, do the multiplication and see which one gives you $3x^2 + 2x - 8$:

$(x+1)(3x-8) = 3x^2 - 5x - 8$	Doesn't work
$(x-1)(3x+8) = 3x^2 + 5x - 8$	Doesn't work
$(3x+1)(x-8) = 3x^2 - 23x - 8$	Doesn't work
$(3x-1)(x+8) = 3x^2 + 23x - 8$	Doesn't work
$(x+4)(3x-2) = 3x^2 + 10x - 8$	Doesn't work
$(x-4)(3x+2) = 3x^2 - 10x - 8$	Doesn't work
$(3x+4)(x-2) = 3x^2 - 2x - 8$	Doesn't work
$(3x-4)(x+2) = 3x^2 + 2x - 8$	Works

Factoring quadratic expressions requires patience and practice. I have given you three more quadratic expressions to factor. The more practice you have, the easier factoring will become.

Lesson 9-4 Practice

Factor the following polynomials:

1. $3x^2 - 10x + 8$ 2. $4x^2 - 16x + 15$ 3. $6x^2 - 29x - 5$

Lesson 9-5: Solving Quadratic Equations by Factoring

The key to solving factored polynomial equations is to use the very special property of 0: whenever a product of terms is 0, you know that at least one of the terms has to equal 0. So if you have a factored polynomial that equals 0, just set each factor equal to 0 and solve. Now we can complicate things a bit. In order to solve a quadratic equation you must first collect all of the terms on one side of the equation, and have 0 on the other side of the equation. Then you factor the quadratic polynomial using the techniques we just discussed. Finally, set each factor equal to 0 and solve. Each one of these examples enables us to put our factoring abilities on the line.

Example 1

Solve the equation: $7x^2 - 10x + 3 = 0$

Solution: Factor the quadratic expression on the left:

$$7x^2 - 10x + 3 = (7x - 3)(x - 1).$$

Next, set each factor equal to 0:

$$(7x - 3)(x - 1) = 0$$

$$(7x - 3) = 0 \qquad\qquad (x - 1) = 0$$
$$7x = 3 \qquad\qquad\qquad x = 1$$
$$x = \frac{3}{7}$$

Our solution is: $x = \dfrac{3}{7}$ or $x = 1$

Example 2

Solve the equation: $10x^2 + 5x - 10 = 2x + 8$

Solution: Collect all of the terms on the left side of the equation and get a 0 on the right:

$10x^2 + 5x - 10 = 2x + 8$

$10x^2 + 3x - 10 = 8$

$10x^2 + 3x - 18 = 0$

Next, factor the quadratic expression:

$10x^2 + 3x - 18 = (5x - 6)(2x + 3)$

Finally, set each factor equal to 0 and solve:

$$(5x - 6)(2x + 3) = 0$$

$(5x - 6) = 0$ $\qquad\qquad$ $(2x + 3) = 0$

$\quad 5x = 6$ $\qquad\qquad\qquad$ $2x = -3$

$\quad x = \dfrac{6}{5}$ $\qquad\qquad\qquad$ $x = -\dfrac{3}{2}$

Our solution is: $x = \dfrac{6}{5}$ or $x = -\dfrac{3}{2}$

Solving quadratic equations provides more opportunities for you to practice your factoring skills. So if you were disappointed by the number of practice problems available in the last section and wanted to factor more, here are some more problems!

Lesson 9-5 Practice

Solve the following quadratic equations:

1. $x^2 - 2x - 8 = 0$ $\qquad\qquad$ 3. $3x^2 + 8x - 3 = 0$

2. $2x^2 - x - 1 = 0$ $\qquad\qquad$ 4. $6x^2 - x - 1 = 0$

Lesson 9-6: Solving Quadratic Equations by Completing the Square

The factoring technique discussed in Lesson 9-5 only works if the linear factors have integer coefficients. In other words, only when the discriminant is a perfect square. We need a technique to solve a quadratic equation when the discriminant is not a perfect square. That technique is completing the square.

Completing the square is a useful little trick for solving quadratic equations. The idea makes use of the technique used to solve $x^2 = c$. Remember that the solution to this type of equation is $x = \pm\sqrt{c}$. We will try to turn our quadratic equations into equations of the form $(x-k)^2 = p$. We can then solve this equation for x:

$$(x-k)^2 = p$$
$$(x-k) = \pm\sqrt{p}$$
$$x = k \pm \sqrt{p}$$

One advantage of this technique is that there is no need to factor!

You need to understand the differences between this technique and the factoring technique discussed in the last lesson. The factoring technique discussed in the last lesson involved comparing a factored quadratic expression to 0. Each factor was then set equal to 0 and solved. Completing the square is one of the rare times when 0 doesn't play a key role. When completing the square, the key is to collect all of the terms that involve variables on one side of the equation, and put all of the constant terms on the other side. We will then make use of what we observed when we squared a binomial:

$$(x+k)^2 = x^2 + 2kx + k^2$$

Notice that the middle term of the trinomial is twice the value of the constant term in the binomial, k. From a different perspective, observe that the constant k in the binomial is one-half of the coefficient of the middle term in the trinomial. Also notice that the constant term of the trinomial is k^2, which is the square of the constant term in the binomial. We will start with the first two terms in the trinomial, and our goal will be to determine the value k.

Let's start with a quadratic equation with leading coefficient equal to 1 and middle term coefficient equal to b, and start moving things around:

$$x^2 + bx + c = 0$$
$$x^2 + bx = -c$$

Next, we need to add a number to both sides so that the expression on the left is a perfect square. Based on our earlier observations, the value of k in the in perfect square is one-half the value of the coefficient of the middle term. In other words, $k = \frac{1}{2}b$. The constant term that we need on the left is k^2, or $\left(\frac{1}{2}b\right)^2 = \frac{1}{4}b^2$. Our expression on the left needs that constant term, so we'll need to add it to both sides of the equation in order to keep the balance:

$$x^2 + bx + \frac{1}{4}b^2 = -c + \frac{1}{4}b^2$$

Now the expression on the left is a perfect square. It has to be, because we added the constant that made it a perfect square. Because it is a perfect square, we can factor it:

$$\left(x + \frac{1}{2}b\right)^2 = -c + \frac{1}{4}b^2$$

From here we know how to solve for x. The general form may look messy, but don't let that intimidate you. Try to focus on how it was derived rather than trying to memorize the final formula.

Example 1

Solve the quadratic equation $x^2 + 4x - 2 = 0$ by completing the square.

Solution: Keep the terms involving variables on one side of the equation and move the constant over to the other side:

$$x^2 + 4x - 2 = 0$$
$$x^2 + 4x = 2$$

Determine what you need to add to both sides: take one-half of the coefficient in front of x and square it: $\left(\frac{1}{2}4\right)^2 = 2^2 = 4$. Add this to both sides of the equation:

$$x^2 + 4x = 2$$

$$x^2 + 4x + 4 = 2 + 4$$

Now, the expression on the left is a perfect square:

$$x^2 + 4x + 4 = 6$$

$$(x+2)^2 = 6$$

Finally, we can solve for x:

$$(x+2)^2 = 6$$

$$(x+2) = \pm\sqrt{6}$$

$$x = -2 + \sqrt{6}$$

Example 2

Solve the quadratic equation $x^2 - 6x - 5 = 0$ by completing the square.

Solution: Keep the terms involving variables on one side of the equation and move the constant over to the other side:

$$x^2 - 6x - 5 = 0$$

$$x^2 - 6x = 5$$

Determine what you need to add to both sides: take one-half of the coefficient in front of x and square it: $\left(\frac{1}{2}6\right)^2 = 3^2 = 9$. It doesn't matter whether the middle coefficient is positive or negative; when you square it you will always end up with a non-negative number. Add this to both sides of the equation:

$$x^2 - 6x = 5$$

$$x^2 - 6x + 9 = 5 + 9$$

Now, the expression on the left is a perfect square:

$$x^2 - 6x + 9 = 14$$

$$(x-3)^2 = 14$$

Finally, we can solve for x:

$$(x-3)^2 = 14$$
$$(x-3) = \pm\sqrt{14}$$
$$x = 3 \pm \sqrt{14}$$

Example 3

Solve the quadratic equation $x^2 + x - 4 = 0$ by completing the square.

Solution: Keep the terms involving variables on one side of the equation and move the constant over to the other side:

$$x^2 + x - 4 = 0$$
$$x^2 + x = 4$$

Determine what you need to add to both sides: take one-half of the coefficient in front of x and square it: $\left(\frac{1}{2} \cdot 1\right)^2 = \left(\frac{1}{2}\right)^2 = \frac{1}{4}$. Add this to both sides of the equation:

$$x^2 + x = 4$$
$$x^2 + x + \frac{1}{4} = 4 + \frac{1}{4}$$

Now, the expression on the left is a perfect square:

$$x^2 + x + \frac{1}{4} = 4 + \frac{1}{4}$$
$$\left(x + \frac{1}{2}\right)^2 = \frac{17}{4}$$

Finally, we can solve for x:

$$\left(x + \frac{1}{2}\right)^2 = \frac{17}{4}$$
$$\left(x + \frac{1}{2}\right) = \pm\sqrt{\frac{17}{4}}$$
$$x = -\frac{1}{2} \pm \sqrt{\frac{17}{4}}$$

Lesson 9-6 Practice

Solve the following quadratic equations by completing the square:

1. $x^2 - 3x - 2 = 0$

2. $x^2 + 4x - 4 = 0$

3. $x^2 + x - 1 = 0$

4. $x^2 - 3x + 1 = 0$

Lesson 9-7: Solving Quadratic Equations Using the Quadratic Formula

The quadratic formula is a formula that will solve quadratic equations very easy. You can solve any solvable quadratic equation by using the quadratic formula. Unfortunately, not all quadratic equations are solvable; some quadratic equations have no solutions. One way to determine whether or not a quadratic equation is solvable is to evaluate the **discriminant**. You remember the discriminant. Given the quadratic equation $ax^2 + bx + c = 0$, the discriminant is the value:

$$b^2 - 4ac$$

The quadratic equation will be solvable as long as the discriminant is *not* a negative number. This will be clear once I write out the quadratic formula.

The quadratic formula is a generalization of completing the square. The quadratic formula allows you to cut to the chase rather than working out all of the details involved in completing the square. The quadratic formula gives the solutions to the quadratic equation $ax^2 + bx + c = 0$. The solutions to this equation are:

$$x = \frac{-b \pm \sqrt{b^2 - 4ac}}{2a}$$

Take a moment to look at this equation, which is called the **quadratic formula**. While I would advise you to memorize it, I also would like you to understand where it came from. First of all, notice that underneath the radical is the expression $b^2 - 4ac$. We have seen this expression before; it is called the discriminant. Recall that if the discriminant is a perfect square then the solutions to the quadratic equation will be rational numbers (or maybe even whole numbers if we are lucky!). The quadratic formula helps you understand why this is true. If the discriminant is a perfect square, then the square root of the discriminant will be a whole number. If b and a are also integers, then x will be a ratio of integers (or a rational number). Also, if the discriminant is a negative number, then using the quadratic formula to solve a quadratic equation will involve taking the square root of a negative number. As we have already discussed, there is no real number that, when squared, is a negative number. We have already established that we will not take the square root of a negative number at this point in our mathematical career.

The motivated reader will want to know where this formula came from. I would answer that this formula is derived by completing the square; if you start with the equation $ax^2 + bx + c = 0$ and complete the square carefully, you will end up deriving the quadratic formula. For now, I will focus on using the formula correctly.

Example 1

Solve the quadratic equation $x^2 + 4x - 2 = 0$ using the quadratic formula.

Solution: In this case, $a = 1$, $b = 4$, and $c = -2$. Using the quadratic formula we have:

$$x = \frac{-b \pm \sqrt{b^2 - 4ac}}{2a}$$

$$x = \frac{-4 \pm \sqrt{4^2 - 4(1)(-2)}}{2(1)}$$

$$x = \frac{-4 \pm \sqrt{16 + 8}}{2}$$

$$x = \frac{-4 \pm \sqrt{24}}{2}$$

Now, we can factor 24: $24 = 4 \cdot 6$, and pull out the perfect square:

$$x = \frac{-4 \pm \sqrt{24}}{2}$$

$$x = \frac{-4 \pm 2\sqrt{6}}{2}$$

$$x = \frac{1}{2}\left(-4 \pm 2\sqrt{6}\right)$$

$$x = -2 \pm \sqrt{6}$$

QUADRATIC EQUATIONS

9

Example 2

Solve the quadratic equation $x^2 - 6x - 5 = 0$ using the quadratic formula.

Solution:

In this case, $a = 1$, $b = -6$, and $c = -5$. Using the quadratic formula carefully we have:

$$x = \frac{-b \pm \sqrt{b^2 - 4ac}}{2a}$$

$$x = \frac{-(-6) \pm \sqrt{(-6)^2 - 4(1)(-5)}}{2(1)}$$

$$x = \frac{6 \pm \sqrt{36 + 20}}{2}$$

$$x = \frac{6 \pm \sqrt{56}}{2}$$

Now, before we move on to the next problem, there is some simplification that we can do. We can factor 56 as $56 = 4 \cdot 14$, and 4 is a perfect square:

$$x = \frac{6 \pm \sqrt{56}}{2}$$

$$x = \frac{6 \pm 2\sqrt{14}}{2}$$

$$x = \frac{1}{2}\left(6 \pm 2\sqrt{14}\right)$$

$$x = 3 \pm \sqrt{14}$$

Example 3

Solve the quadratic equation $2x^2 - x - 4 = 0$ using the quadratic formula.

Solution:

In this case, $a = 2$, $b = -1$, and $c = -4$. Using the quadratic formula carefully we have:

$$x = \frac{-b \pm \sqrt{b^2 - 4ac}}{2a}$$

$$x = \frac{-(-1) \pm \sqrt{(-1)^2 - 4(2)(-4)}}{2(2)}$$

$$x = \frac{1 \pm \sqrt{1 + 32}}{4}$$

$$x = \frac{1 \pm \sqrt{33}}{4}$$

Because 33 has no factors that are perfect squares, our work here is done.

Lesson 9-7 Practice

Solve the following quadratic equations using the quadratic formula:

1. $x^2 - 2x - 1 = 0$

2. $x^2 + 5x + 6 = 0$

3. $2x^2 - 3x - 7 = 0$

4. $3x^2 + 2x - 5 = 0$

9

QUADRATIC EQUATIONS

QUADRATIC EQUATIONS

9

Answer Key
Lesson 9-1

1. a. $3\sqrt{7}$

   b. $4\sqrt{5}$

   c. $6\sqrt{3}$

2. $\sqrt{265}$

Lesson 9-2a

1. $(x-6)(x+6)$
2. $\left(x-\sqrt{15}\right)\left(x+\sqrt{15}\right)$
3. $(3x-2)(3x+2)$

Lesson 9-2b

1. $x=\pm 9$

2. $x=\pm 3\sqrt{3}$

3. $x=0$ or $x=-4$

4. $x=2\pm\sqrt{10}$

Lesson 9-3

1. a. $(x-2)(x+1)$

   b. $(x+6)(x-1)$

   c. $(x+4)(x+1)$

   d. $(x-2)(x-4)$

2. a. The discriminant is 25 which is a perfect square, so the linear factors will have integer coefficients.

   b. The discriminant is 17 which is not a perfect square, so the linear factors will not have integer coefficients.

Lesson 9-4

1. $(3x-4)(x-2)$

2. $(2x-5)(2x-3)$

3. $(6x+1)(x-5)$

Lesson 9-5

1. $(x-4)(x+2)=0$; $x=4$ or $x=-2$

2. $(2x+1)(x-1)=0$; $x=-\dfrac{1}{2}$ or $x=1$

3. $(3x-1)(x+3)=0$; $x=\dfrac{1}{3}$ or $x=-3$

4. $(3x+1)(2x-1)=0$; $x=-\dfrac{1}{3}$ or $x=\dfrac{1}{2}$

Lesson 9-6

1. $x^2-3x=2$

$$x^2-3x+\frac{9}{4}=2+\frac{9}{4}$$

$$\left(x-\frac{3}{2}\right)^2=\frac{17}{4}$$

$$x=\frac{3}{2}\pm\frac{\sqrt{17}}{2}$$

2. $x^2+4x=4$

$$x^2+4x+4=4+4$$

$$(x+2)^2=8$$

$$x=-2\pm2\sqrt{2}$$

3. $x^2+x=1$

$$x^2+x+\frac{1}{4}=1+\frac{1}{4}$$

$$\left(x+\frac{1}{2}\right)^2=\frac{5}{4}$$

$$x = -\frac{1}{2} \pm \frac{\sqrt{5}}{2}$$

4. $x^2 - 3x = -1$

$$x^2 - 3x + \frac{9}{4} = -1 + \frac{9}{4}$$

$$\left(x - \frac{3}{2}\right)^2 = \frac{5}{4}$$

$$x = \frac{3}{2} \pm \frac{\sqrt{5}}{2}$$

Lesson 9-7

1. $x^2 - 2x - 1 = 0$

$$x = \frac{2 \pm \sqrt{4 - 4(1)(-1)}}{2 \cdot 1}$$

$$x = 1 \pm \sqrt{2}$$

2. $x^2 + 5x + 6 = 0$

$$x = \frac{-5 \pm \sqrt{25 - 4(1)(6)}}{2 \cdot 1}$$

$x = -3$ or $x = -2$

3. $2x^2 - 3x - 7 = 0$

$$x = \frac{3 \pm \sqrt{9 - 4(2)(-7)}}{2 \cdot 2}$$

$$x = \frac{3 \pm \sqrt{65}}{4}$$

4. $3x^2 + 2x - 5 = 0$

$$x = \frac{-2 \pm \sqrt{4 - 4(3)(-5)}}{2 \cdot 3}$$

$x = -\frac{5}{3}$ or $x = 1$

10

Rational Expressions

When thinking of things we can do with *functions*, we look at the things we can do with *numbers*. Using the integers as our basic building block, other kinds of numbers were created. Taking *ratios* of two integers enabled us to create the *rational* numbers. If we use polynomials as our basic building blocks in algebra, we can take the ratio of two *polynomials* to create a new function, called a **rational expression**, or a rational function. The rules for manipulating rational expressions will be directly related to the rules for manipulating fractions. You should draw from your understanding of fractions (or rational numbers) as we study rational expressions.

Lesson 10-1: Simplifying Rational Expressions

When you see a fraction like $\frac{10}{15}$ the first thought that you have may be that this fraction is not in reduced form. A fraction is in reduced form if the numerator and the denominator do not have any factors in common. This is the first idea that we will make use of with rational expressions.

10 RATIONAL EXPRESSIONS

A rational expression may or may not be in reduced form. Your familiarity with numbers helped you determine whether a fraction was written in reduced form. But if a fraction involves large numbers that you are not familiar with, like the fraction $\frac{2{,}491}{9{,}741}$, you may find yourself creating factor trees for the numerator and the denominator in order to see if there are any common factors. The same holds true for rational expressions. The factors of a polynomial don't always jump out at you. It will take some work, by way of factoring, to determine whether or not a rational expression is in reduced form.

To see whether a rational expression is in reduced form or not, the first thing you need to do is factor the polynomials in the numerator and the denominator. This will help you find any common factors that can then be cancelled. The instruction for writing a rational expression in reduced form is "simplify." When you see the instruction "simplify" followed by a rational expression, you should immediately address whether or not the rational expression is in reduced form. And to do that, you will need to factor both polynomials involved in the rational expression.

Example 1

Simplify the expression: $\dfrac{x^2 + x - 6}{x^2 - x - 12}$

Solution: Factor the numerator and denominator using the techniques discussed in the last chapter and then cancel any common factors:

$$\frac{x^2 + x - 6}{x^2 - x - 12} = \frac{(x-2)(x+3)}{(x-4)(x+3)} = \frac{(x-2)}{(x-4)}$$

You may be tempted to try to cancel other things in this fraction. Be very careful when you cancel terms. You are only allowed to cancel *factors*; you can only cancel terms that are being ***multiplied together***.

You **cannot** cancel across an addition or subtraction sign. Even though there is a 2 in the numerator and a 4 in the denominator, they are being subtracted, not multiplied. You need to think of linear terms like $(x - 2)$ as a unit...the 2 is attached to the x through subtraction, and any terms that are connected by subtraction (or addition) cannot be canceled by themselves. Putting parentheses around any terms that involve addition or subtraction will help remind you that you must treat those terms as a group, and you cannot cancel individual terms in a group.

Example 2

Simplify the expression: $\dfrac{x^2 - 5x - 6}{x^2 - 3x - 4}$

Solution: Factor the numerator and denominator using the techniques discussed in the last chapter and then cancel any common factors:

$$\frac{x^2 - 5x - 6}{x^2 - 3x - 4} = \frac{(x-6)(x+1)}{(x-4)(x+1)} = \frac{(x-6)}{(x-4)}$$

An important step in simplifying rational expressions is to factor the numerators and denominators correctly. It is very important that you master factoring.

Lesson 10-1 Practice

Simplify the following rational expressions:

1. $\dfrac{x^2 - x - 6}{x^2 - 2x - 3}$

2. $\dfrac{x^2 - 3x - 4}{x^2 + 3x + 2}$

3. $\dfrac{x^2 + 2x - 8}{x^2 - x - 2}$

10 QUADRATIC EQUATIONS

Lesson 10-2: Multiplying Rational Expressions

Multiplying rational expressions is a lot like multiplying fractions. When you multiply two fractions you multiply the numerators together to get the new numerator and you multiply the denominators together to get the new denominator. We will do the same thing when multiplying rational expressions. To keep our expressions simple, we will factor and cancel wherever possible. Keep your final answer in factored form.

Example 1

Find the product: $\dfrac{x-1}{x^2-x-2} \cdot \dfrac{x-2}{x^2-x}$

Solution: Before you start multiplying numerators and denominators together, I recommend factoring every polynomial. Use parentheses to help keep the terms grouped together where they belong. Once everything is factored you can cancel any common factors:

$$\frac{x-1}{x^2-x-2} \cdot \frac{x-2}{x^2-x} = \frac{\cancel{(x-1)}\cancel{(x-2)}}{\cancel{(x-2)}(x+1)(x)\cancel{(x-1)}} = \frac{1}{x(x+1)}$$

Example 2

Find the product: $\dfrac{x^2+8x+12}{x^2-5x-6} \cdot \dfrac{x+1}{x^2+5x-6}$

Solution: Factor each polynomial and cancel what you can. Don't be afraid to use parentheses to organize the factors:

$$\frac{x^2+8x+12}{x^2-5x-6} \cdot \frac{x+1}{x^2+5x-6} = \frac{\cancel{(x+6)}(x+2)\cancel{(x+1)}}{(x-6)\cancel{(x+1)}\cancel{(x+6)}(x-1)}$$

$$= \frac{(x+2)}{(x-6)(x-1)}$$

Lesson 10-2 Practice

Find the following products:

1. $\dfrac{x^2+x-2}{x^2-9} \cdot \dfrac{x-3}{x-1}$

2. $\dfrac{x^2-4}{x^2-16}\cdot\dfrac{x-4}{x-2}$

3. $\dfrac{x+2}{x^2+x-6}\cdot\dfrac{x+3}{x-1}$

4. $\dfrac{x-2}{x^2+5x+4}\cdot\dfrac{x+4}{x-1}$

Lesson 10-3: Dividing Rational Expressions

Fractions are the motivation for manipulating rational expressions. Dividing one rational expression by another rational expression will follow the same rules established for dividing one fraction by another fraction. When you divide one fraction by another fraction, the fraction that is in the denominator (or the fraction that is the divisor) is inverted and the resulting two fractions are then multiplied.

There are several ways to write the ratio of two fractions. Sometimes one format is easier to read than the others. I will mention two specific formats that represent the same ratio.

- In the ratio $\dfrac{\frac{a}{b}}{\frac{c}{d}}$, the numerator is $\dfrac{a}{b}$ and the denominator is $\dfrac{c}{d}$

- In the ratio $\dfrac{a}{b} \, \textrm{|} \, \dfrac{c}{d}$, the numerator is $\dfrac{a}{b}$ and the denominator is $\dfrac{c}{d}$

The second way is a bit easier to read than the first way, and that will be the format of choice in this section.

Because rational expressions behave like fractions, it should not surprise you to learn that when you divide one rational expression by another rational expression you will invert the rational expression in the denominator and multiply.

Example 1

Simplify the expression: $\dfrac{x^2+x-2}{x+5} \div \dfrac{x+2}{x-2}$

Solution: The key is to invert the denominator, which in this case is the expression $\dfrac{x+2}{x-2}$, and then multiply. After you have inverted the denominator and are getting ready to multiply, you will want to factor all of the polynomials and cancel any common factors:

$$\dfrac{x^2+x-2}{x+5} \div \dfrac{x+2}{x-2}$$

Invert the denominator $\dfrac{x+2}{x-2}$ and multiply:

$$\dfrac{x^2+x-2}{x+5} \cdot \dfrac{x-2}{x+2}$$

Factor the quadratic expression x^2+x-2:

$$\dfrac{\cancel{(x+2)}(x-1)(x-2)}{(x+5)\cancel{(x+2)}}$$

Cancel common linear factors:

$$\dfrac{(x-1)(x-2)}{(x+5)}$$

Example 2

Simplify the expression: $\dfrac{x^2+4x+3}{x-3} \div \dfrac{x^2-9}{x-1}$

Solution: Invert, multiply, factor, and cancel:

$$\dfrac{x^2+4x+3}{x-3} \div \dfrac{x^2-9}{x-1}$$

Invert the denominator $\dfrac{x^2-9}{x-1}$ and multiply:

$$\dfrac{x^2+4x+3}{x-3} \cdot \dfrac{x-1}{x^2-9}$$

Factor the quadratic expressions x^2+4x+3 and x^2-9:

$$\dfrac{\cancel{(x+3)}(x+1)(x-1)}{(x-3)(x-3)\cancel{(x+3)}}$$

Cancel common linear factors:

$$\dfrac{(x+1)(x-1)}{(x-3)^2}$$

Lesson 10-3 Practice

Simplify the following expressions:

1. $\dfrac{x^2-1}{x+2} \div \dfrac{x-1}{x^2-4}$

2. $\dfrac{x+1}{x-2} \div \dfrac{x^2-9}{x-2}$

3. $\dfrac{x^2+x-6}{x+1} \div \dfrac{x-2}{x+1}$

4. $\dfrac{x^2+6x+5}{x-2} \div \dfrac{x+1}{x^2-3x+2}$

Lesson 10-4: Adding and Subtracting Rational Expressions

Adding and subtracting fractions was more involved than multiplying and dividing fractions. Similarly, adding and subtracting rational expressions is a bit more involved than multiplying or dividing rational expressions. Recall that you can only add two fractions if they have a *common denominator*. If the denominators are the same, then you are all set. But if the denominators are different then you have to find a common denominator. The best common denominator is the *least common denominator*. In order to find the least common denominator you must factor both denominators completely and look for common factors.

Example 1

Simplify: $\dfrac{x+1}{x-2} + \dfrac{3x+4}{x-2}$

Solution: The denominators are the same, so all you have to do is add the numerators. Remember the rules for adding polynomials. Only like terms can be added (or subtracted) from each other:

$$\frac{x+1}{x-2} + \frac{3x+4}{x-2} = \frac{4x+5}{x-2}$$

10 QUADRATIC EQUATIONS

Example 2

Simplify: $\dfrac{x+1}{x-2} + \dfrac{x+4}{x+3}$

Solution: Notice that neither denominator can be factored. Therefore, the least common denominator is found by multiplying the terms in the denominator together: $(x-2)(x+3)$. Then, remember, that to get the two denominators to match we can multiply by the number 1 disguised using the term that is needed in each denominator. The first term in the expression is $\dfrac{x+1}{x-2}$; it should be multiplied by $\dfrac{x+3}{x+3}$. The second term in the expression is $\dfrac{x+4}{x+3}$; it should be multiplied by $\dfrac{x-2}{x-2}$. Once the denominators are the same you will need to expand the numerators and perform the addition:

$$\frac{x+1}{x-2} + \frac{x+4}{x+3}$$

Multiply the first expression by $\dfrac{x+3}{x+3}$ and the second expression by $\dfrac{x-2}{x-2}$:

$$\frac{x+3}{x+3} \cdot \frac{x+1}{x-2} + \frac{x+4}{x+3} \cdot \frac{x-2}{x-2}$$

$$\frac{(x+3)(x+1)}{(x+3)(x-2)} + \frac{(x+4)(x-2)}{(x+3)(x-2)}$$

Add the two fractions together:

$$\frac{(x+3)(x+1)+(x+4)(x-2)}{(x+3)(x-2)}$$

Expand each product in the numerator:

$$\frac{\left(x^2+4x+3\right)+\left(x^2+2x-8\right)}{(x+3)(x-2)}$$

Add the two polynomials together:

$$\frac{2x^2+6x-5}{(x+3)(x-2)}$$

Example 3

Simplify: $\dfrac{x+2}{x-1} - \dfrac{x+3}{x+2}$

Solution: Neither denominator can be factored, so we'll follow the same approach as we did in Example 2.

$$\frac{x+2}{x-1} - \frac{x+3}{x+2}$$

Multiply the first expression by $\dfrac{x+2}{x+2}$

and the second expression by $\dfrac{x-1}{x-1}$:

$$\frac{x+2}{x+2} \cdot \frac{x+2}{x-1} - \frac{x+3}{x+2} \cdot \frac{x-1}{x-1}$$

$$\frac{(x+2)(x+2)}{(x+2)(x-1)} - \frac{(x+3)(x-1)}{(x+2)(x-1)}$$

Subtract the two fractions:

$$\frac{(x+2)(x+2)-(x+3)(x-1)}{(x+2)(x-1)}$$

Expand each product in the numerator:

$$\frac{\left(x^2+4x+4\right)-\left(x^2+2x-3\right)}{(x+2)(x-1)}$$

Subtract the polynomials:

$$\frac{2x+7}{(x+2)(x-1)}$$

Before continuing to the next example, I would like to remind you of how to calculate the least common multiple of two numbers. One way that we calculated the least common multiple was to take the product of the two numbers and divide by the greatest common factor. To find the least common multiple of 15 and 25, we would first factor each number to find the greatest common factor: $15 = 3 \times 5$ and $25 = 5 \times 5$. The greatest common factor of 15 and 25 is 5. To find the least common multiple, evaluate $\frac{15 \times 25}{5} = \frac{3 \times \cancel{5} \times 5 \times 5}{\cancel{5}} = 75$. The least common multiple of 15 and 25 is 75. We can apply this technique to finding the least common multiple of two polynomials, as we will see in the next example.

Example 4

Simplify: $\dfrac{3}{4x} + \dfrac{2}{6x^2}$

Solution: In this case, notice that the denominators have factors in common. To find the least common denominator, factor each

10 QUADRATIC EQUATIONS

denominator completely: $4x = 2 \cdot 2 \cdot x$ and $6x^2 = 2 \cdot 3 \cdot x \cdot x$. Notice that the two denominators have some factors in common: they both have a factor of $2x$. To find the least common multiple of $4x$ and $6x^2$, evaluate the ratio $\dfrac{(4x)(6x^2)}{2x}$:

$$\frac{(4x)(6x^2)}{2x} = \frac{24x^3}{2x} = \left(\frac{24}{2}\right)\left(\frac{x^3}{x}\right) = 12x^2$$

The least common multiple of $4x$ and $6x^2$ is $12x^2$.

Now let's return to the problem of finding the sum $\dfrac{3}{4x} + \dfrac{2}{6x^2}$.

In order to make the denominator of the first expression $12x^2$ we will need to multiply the first expression by $\dfrac{3x}{3x}$.

In order to make the denominator of the second expression $12x^2$ we will need to multiply the second expression by $\dfrac{2}{2}$:

$$\frac{3}{4x} + \frac{2}{6x^2}$$

Multiply the first expression by $\dfrac{3x}{3x}$

and the second expression by $\dfrac{2}{2}$: $\quad \dfrac{3x}{3x} \cdot \dfrac{3}{4x} + \dfrac{2}{6x^2} \cdot \dfrac{2}{2}$

Find each product: $\quad \dfrac{9x}{12x^2} + \dfrac{4}{12x^2}$

Add the two numerators: $\quad \dfrac{9x+4}{12x^2}$

Lesson 10-4 Practice

Simplify the following expressions:

1. $\dfrac{3}{x+1} + \dfrac{4}{x-2}$

2. $\dfrac{x}{x-1} - \dfrac{3}{x+1}$

3. $\dfrac{x+2}{x+4} + \dfrac{x-1}{x+1}$

4. $\dfrac{x+5}{x+6} - \dfrac{x}{2x+1}$

Answer Key
Lesson 10-1

1. $\dfrac{(x-3)(x+2)}{(x-3)(x+1)}=\dfrac{x+2}{x+1}$

2. $\dfrac{(x-4)(x+1)}{(x+2)(x+1)}=\dfrac{x-4}{x+2}$

3. $\dfrac{(x+4)(x-2)}{(x+1)(x-2)}=\dfrac{x+4}{x+1}$

Lesson 10-2

1. $\dfrac{(x+2)(x-1)}{(x+3)(x-3)}\cdot\dfrac{(x-3)}{(x-1)}=\dfrac{x+2}{x+3}$

2. $\dfrac{(x+2)(x-2)}{(x+4)(x-4)}\cdot\dfrac{(x-4)}{(x-2)}=\dfrac{x+2}{x+4}$

3. $\dfrac{(x+2)}{(x+3)(x-2)}\cdot\dfrac{(x+3)}{(x-1)}=\dfrac{x+2}{(x-2)(x-1)}$

4. $\dfrac{(x-2)}{(x+4)(x+1)}\cdot\dfrac{(x+4)}{(x-1)}=\dfrac{x-2}{(x+1)(x-1)}$

Lesson 10-3

1. $\dfrac{x^2-1}{x+2}\div\dfrac{x-1}{x^2-4}=\dfrac{x^2-1}{x+2}\cdot\dfrac{x^2-4}{x-1}=\dfrac{(x-1)(x+1)}{(x+2)}\cdot\dfrac{(x-2)(x+2)}{(x-1)}$

$=(x+1)(x-2)$

2. $\dfrac{x+1}{x-2}\div\dfrac{x^2-9}{x-2}=\dfrac{x+1}{x-2}\cdot\dfrac{x-2}{x^2-9}=\dfrac{(x+1)}{(x-2)}\cdot\dfrac{(x-2)}{(x+3)(x-3)}$

$=\dfrac{x+1}{(x-3)(x+3)}$

3. $\dfrac{x^2+x-6}{x+1}\div\dfrac{x-2}{x+1}=\dfrac{x^2+x-6}{x+1}\cdot\dfrac{x+1}{x-2}=\dfrac{(x+3)(x-2)}{(x+1)}\cdot\dfrac{(x+1)}{(x-2)}$

$=(x+3)$

$$\frac{x^2+6x+5}{x-2} \div \frac{x+1}{x^2-3x+2} = \frac{x^2+6x+5}{x-2} \cdot \frac{x^2-3x+2}{x+1}$$

4.
$$= \frac{(x+5)(x+1)}{(x-2)} \cdot \frac{(x-2)(x-1)}{(x+1)}$$

$$= (x+5)(x-1)$$

Lesson 10-4

1. $\dfrac{3}{x+1} + \dfrac{4}{x-2} = \dfrac{(x-2)}{(x-2)} \cdot \dfrac{3}{(x+1)} + \dfrac{4}{(x-2)} \cdot \dfrac{(x+1)}{(x+1)} = \dfrac{7x-2}{(x+1)(x-2)}$

2. $\dfrac{x}{x-1} - \dfrac{3}{x+1} = \dfrac{(x+1)}{(x+1)} \cdot \dfrac{x}{(x-1)} - \dfrac{3}{(x+1)} \cdot \dfrac{(x-1)}{(x-1)} = \dfrac{x^2-2x+3}{(x-1)(x+1)}$

3. $\dfrac{x+2}{x+4} + \dfrac{x-1}{x+1} = \dfrac{(x+1)}{(x+1)} \cdot \dfrac{(x+2)}{(x+4)} + \dfrac{(x-1)}{(x+1)} \cdot \dfrac{(x+4)}{(x+4)} = \dfrac{2x^2+6x-2}{(x+4)(x+1)}$

4. $\dfrac{x+5}{x+6} - \dfrac{x}{2x+1} = \dfrac{(2x+1)}{(2x+1)} \cdot \dfrac{(x+5)}{(x+6)} - \dfrac{x}{(2x+1)} \cdot \dfrac{(x+6)}{(x+6)}$

$$= \frac{x^2+5x+5}{(x+6)(2x+1)}$$

11

Applications

The reason that algebra has been around for so long is because it is so useful: it provides a systematic way to solve problems. Until now, I have given you equations and asked you to solve them. In this chapter, the tables will be turned. I will describe a problem in words and you will get to come up with the mathematical equation to solve. You will also have to solve the equation that you create.

Lesson 11-1: How to Approach Word Problems

A **word problem** involves a description of a situation, and a request for an answer to a question. There are no variables in the word problem; it is your job to come up with the variables, and to write an equation based on the information given in the problem. You could generate one equation or a system of equations. Once you have an equation (or equations), you will need to solve it (or them) and then use your solution (or solutions) to answer the question asked in the problem.

Some of the equations that you create will be linear, others could be quadratic. If you generate a quadratic equation you will either need to factor it or use the quadratic formula to solve it.

I recommend using a systematic method for approaching and solving word problems. Here is one problem-solving approach that will help you solve word problems.

- ▣ Read the problem description and pick out the important information. Determine what it is that you are given and what you are asked to find. Choose variables to represent what you are given and what you need to find. You are creating the equations, so you can choose the variables. Don't limit yourself to the variables x and y; sometimes it's helpful to use the first letter of what the variable represents. If you get too deep into a problem it's easy to forget what your variables mean, so it's a good idea to make a note of it somewhere.

- ▣ Read the word problem and interpret the description of the problem in terms of your variables. Be sure to pay attention to the units involved. There are many ways to represent addition, subtraction, multiplication, and division. Words such as "sum," "more," "and," "total," "plus," or "increase" indicate that terms should be *added*. Words like "difference," "less than," "fewer," or "decreased" indicate that terms should be *subtracted*. *Multiplication* is often referred to using words such as "product," "times," "twice," "percent" or "of." *Division* will be involved when you see words like "ratio," "divided by," "half," or "quotient."

- ▣ Solve the equation or evaluate the expression for a particular value of one of your variables. Use the problem-solving strategies discussed in the earlier chapters to help you solve equations. Be sure to check your answers to catch any mistakes.

▣ Re-read the question and be sure that the answer you found actually addresses the question asked. **You should think about your answer to make sure that it makes sense.**

If you use this approach to solve word problems, you should have great success solving word problems, and as much fun solving them as I have.

Lesson 11-2: Rates and Percentages

A **rate** is a ratio of two quantities that are measured in different units. For example, the rate that you drive on some interstates is 60 miles per hour. This is a ratio of distance (miles) divided by time (hours). A **unit rate** is a rate where the denominator of the ratio is 1. You can think of the rate 60 miles per hour as 60 miles per 1 hour. A **proportion** is an equality between two rates. Proportions can be used to solve many problems. The algebraic step necessary to solve many proportions is to **cross-multiply** to clear out the fractions:

$$\text{If } \frac{a}{b} = \frac{c}{d}, \text{ then } a \cdot d = b \cdot c$$

Example 1

At the local print shop, 15 copies cost $0.50. At this rate, how much would it cost to make 120 copies?

Solution: Set up a proportion. Let c represent the cost of 120 copies. Then c is to 120 as 0.5 is to 15: $\frac{c}{120} = \frac{0.5}{15}$. Cross-multiply and solve for c:

$$\frac{c}{120} = \frac{0.5}{15}$$

Cross-multiply: $\qquad\qquad 15c = (120)(0.5)$

Simplify: $\qquad\qquad 15c = 60$

Divide both sides by 15: $\qquad \frac{15c}{15} = \frac{60}{15}$

Simplify: $\qquad\qquad c = 4$

Interpret your answer: It would cost $4.00 for 120 copies.

Example 2

The current exchange rate between dollars and euros is 1 Euro = $1.20. If Nathan is traveling in Europe and wants to buy his sister a t-shirt that costs 10 Euros, how much does the shirt cost in dollars?

Solution: Let e represent the price, in *euros*, and d represent the price, in *dollars*. Set up a proportion relating these two currencies:

$$\frac{d}{e} = \frac{\$1.2}{1 \text{ euro}}$$

Substitute in the price of the shirt in euros and solve for the price in dollars by cross-multiplying:

$$\frac{d}{10} = \frac{1.2}{1}$$

$$1 \cdot d = 10 \cdot 1.2$$

$$d = 12$$

Interpret your answer: d represented the price, in dollars, so the answer to the question is that the t-shirt costs $12.

A percent means "out of 100." Decimals can be written as percents and percents can be written as decimals. To convert from a decimal to a percent, multiply the decimal by 100; this is the same thing as moving the decimal two places to the right. To convert from a percent to a decimal divide by 100, or move the decimal two places to the left. As you'll see in this example, percents and proportions go hand in hand.

Example 3

Suppose that in an algebra class 15% of the students will earn an "A" in the class. There are 40 students in the class. How many of them will earn an "A" in the class?

Solution: Let s represent the number of students in the class and let a represent the number of students who will earn an "A" in the class. Set up a proportion:

$$\frac{a}{s} = \frac{15}{100}$$

Replace s by the number of students in the class (40) and solve for a:

$$\frac{a}{40} = \frac{15}{100}$$

$$a \cdot 100 = 15 \cdot 40$$

$$100a = 600$$

$$\frac{\cancel{100}a}{\cancel{100}} = \frac{600}{100}$$

$$a = 6$$

Since a represents the number of students who earn an "A" in the class, there are 6 students who will earn an "A" in the algebra class.

Lesson 11-2 Practice

1. Adam has designed a hybrid car that can travel 55 miles per gallon of gasoline. If the gas tank holds 12 gallons of gasoline, how far can the car travel on one tank of gas?

2. Alan works for a roofing company. Each "square" represents 100 square feet of roof. Alan is paid $80 per square to shingle a roof. If a roof requires 44 squares, how much will Alan get paid to shingle the roof?

3. Julia has invested in the stock market and must now pay taxes on her capital gains. If the IRS taxes Julia at a rate of 22%, how much will Julia owe on $30,000 in capital gains?

Lesson 11-3: Finding Integers

In these types of problems you are given clues about the relationships between two integers, and you have to find the integers.

Example 1

The sum of two integers is 4 and their difference is 20. Find the two integers.

11 **APPLICATIONS**

Solution: Let x represent one integer and y represent the other one. Since their sum is 4 we know that $x + y = 4$. The fact that their difference is 20 means that $x - y = 20$. Now we have a system of two equations and two unknowns that we can solve:

$$\begin{cases} x + y = 4 \\ x - y = 20 \end{cases}$$

We can solve this system using the substitution method: use the second equation to solve for x:

$$x - y = 20$$
$$x = 20 + y$$

Now substitute this expression for x into the first equation:

$$x + y = 4$$

Substitute the value $20 + y$ in for x: $\qquad (20 + y) + y = 4$

Simplify: $\qquad 2y + 20 = 4$

Subtract 20 from both sides: $\qquad 2y = -16$

Divide both sides by 2: $\qquad y = -8$

Now that you know the value of y, you can find x:

$$x = 20 + y$$
$$x = 20 + (-8)$$
$$x = 12$$

So the two integers are 12 and –8. Check to make sure that this answer makes sense: their sum [12 + (–8)] is 4 and their difference [12 – (–8)] is 20.

Example 2

The sum of two integers is 15. One integer is three more than twice the other integer. Find the two integers.

Solution: Let x represent one of the integers and y represent the other integer. Then their sum is 15 and we have our first equation:

$$x + y = 15$$

The second equation is a bit more tricky. One of the integers has to be three more than twice the other. So we will need to double one of our numbers and then add 3 to get the other one:

$x = 2y + 3$

Once again we have a system of equations that we can solve:

$$\begin{cases} x + y = 15 \\ x = 2y + 3 \end{cases}$$

Again, use the elimination method to solve for x and y. Our second equation is ready to substitute into the first equation:

$$x + y = 15$$

Substitute the value $2y + 3$ in for x: $(2y + 3) + y = 15$

Simplify: $3y + 3 = 15$

Subtract 3 from both sides: $3y = 12$

Divide both sides by 3: $y = 4$

Now that we know y,
we can find x: $x = 2y + 3$

$$x = 2(4) + 3$$

$$x = 11$$

So the two integers are 4 and 11.
Check to see if this answer makes sense:
their sum $(4 + 11)$ is 15 and 11 is 3 more than two times 4.
$(11 = 3 + 2 \cdot 4)$

Lesson 11-3 Practice

1. The sum of two integers is 65 and their difference is 25. Find the two integers.

2. The sum of two integers is 40. If one integer is 5 less than twice the other, find the two integers.

11 APPLICATIONS

Lesson 11-4: Rate, Time, and Distance Problems

Rate problems revolve around the idea that rate times time equals distance. If r represents the rate, t is the time and d represents the distance, then this equation can be written:

$$r \cdot t = d$$

Example 1

April plans to drive 595 miles to Tallahassee, Florida, to watch the Florida State Seminoles play the Clemson Tigers. If she drives at a rate of 70 miles per hour, how long will it take her to make the trip?

Solution: Start with the equation $r \cdot t = d$. We know the distance (595 miles) and the rate (70 miles per hour) so we can easily find the time:

$r \cdot t = d$

$70 \cdot t = 595$

$t = \dfrac{595}{70} = \dfrac{\cancel{5} \cdot \cancel{7} \cdot 17}{2 \cdot \cancel{5} \cdot \cancel{7}} = \dfrac{17}{2} = 8.5$

It will take her 8.5 hours to make the trip.

Example 2

Debra is training to enter a triathlon which consists of a 3 mile run, a one-half mile swim and a 15 mile bicycle ride. Debra can run at a rate of 10 miles per hour, she can swim at a rate of $\frac{1}{5}$ of a mile per hour, and she can bike at a rate of 25 miles per hour. How long should it take her to complete the race?

Solution: In this problem we have several rates and corresponding distances for each leg of the race. We need to find the time it takes to run, swim and bike each part of the race and then add these individual times together to get the total time. In order to find the individual times, we will need to solve the rate equation for t (time):

$$r \cdot t = d$$

$$t = \frac{d}{r}$$

Now find the individual times by substituting into the equation:

Running: $\qquad t = \dfrac{d}{r} = \dfrac{3}{10}$

Swimming: $\qquad t = \dfrac{d}{r} = \dfrac{\frac{1}{2}}{\frac{1}{5}} = \dfrac{1}{2} \cdot \dfrac{5}{1} = \dfrac{5}{2}$

Biking: $\qquad t = \dfrac{15}{25} = \dfrac{3}{5}$

Because each calculation resulted in the same units for time, we can just add up the individual times to get the total time. Be sure to pay attention to the units given. It is really easy to slip in a rate with units of miles per minute, in which case the time you calculated would have been measured in minutes instead of hours. Add up the individual times to get the total time to finish the race:

$$t = \frac{3}{10} + \frac{5}{2} + \frac{3}{5} = \frac{3}{10} + \frac{25}{10} + \frac{6}{10} = \frac{34}{10} = 3.4$$

It will take Debra 3.4 hours to complete the race.

Lesson 11-4 Practice

1. Steven is driving at a rate of 60 miles per hour. How many hours will it take him drive 300 miles?

2. Cathy must be at work at 8 a.m. She lives 35 miles from her office and is running late. If she leaves her house at 7:18, how fast, will she have to drive in order to get to work on time? Give her rate in miles per hour.

3. Ken likes to shop at the outlet malls in St. Augustine. Unfortunately, the malls are 150 miles from where he lives. He decides to make the trip to the mall in one day. He leaves his house at 9 a.m. and would like to be home by 8 a.m. He drives

60 miles per hour, and doesn't stop driving until he reaches his destination. How many hours will he be able to spend shopping?

4. Mike is flying from Orlando, Florida to Green Bay, Wisconsin. His estimated time of arrival is 6 a.m. Nicole is planning to pick him up at the airport. She lives 44 miles from the airport and would like to be there on time to pick him up. She leaves her house at 5:10 and gets stuck in rush hour traffic. If she can only drive 48 miles per hour, at what time will she get to the airport?

Lesson 11-5: Money Problems

Problems that involve money usually involve either different coins or bills. If you always work in dollars (when the problems involve bills) or cents (if the problem involves coins), you'll end up dealing with whole numbers instead of fractions or decimals.

Example 1

Tony's parents give him money for getting good grades on his report card. His parents give him $10 for every A and $5 for every B. If Tony earned either an A or a B in each of his 7 classes and his parents gave him $60, how many A's did Tony have on his report card?

Solution: Let a represent the number of As on Tony's report card, and let b represent the number of Bs on his report card. Tony earned only As and Bs on his report card so we know that $a+b=7$. Each A is worth $10, so he earned $10a$ for his As; each B is worth $5, so he earned $5b$ for his Bs. The total amount of money he earned must satisfy the equation $10a+5b=60$. Now we have a system of 2 equations that we can solve:

$$\begin{cases} a+b=7 \\ 10a+5b=60 \end{cases}$$

Solve this equation using the substitution/elimination method:

$$a + b = 7$$

Use this equation to solve for b: $\quad b = 7 - a$

Substitute $7 - a$ in for b in the
second equation: $\qquad 10a + 5(7 - a) = 60$

Distribute the 5: $\qquad 10a + 35 - 5a = 60$

Simplify: $\qquad 5a + 35 = 60$

Subtract 35 from both sides: $\qquad 5a = 25$

Divide both sides by 5: $\qquad a = 5$

Tony earned 5 As on his report card.

Example 2

Betty always gives her grandchildren money for their
birthday. The amount she gives them depends on their age.
If she gives them $10 plus $5 for every year over 13, how
much money will she give her granddaughter, Emily, on her
sixteenth birthday?

Solution: Let G represent the amount of the gift, and let a represent
Emily's age. Then the amount of the birthday gift is given by the
equation:

$$G = 10 + 5(a - 13)$$

Because Emily will be turning 16, so we will need to evaluate the
expression on the right when $a = 16$:

$$G = 10 + 5(16 - 13) = 10 + 5 \cdot 3 = 25$$

Emily will receive $25 from her grandmother on her 16th birthday.

Lesson 11-5 Practice

1. Alic has twice as many ten-dollar bills as five-dollar bills. If he has
   $200, how many five-dollar bills does he have?

2. A cell phone plan costs $40 per month for 300 minutes and $0.35

11

APPLICATIONS

per minute for any overage time. Find the amount of the cell phone bill if a customer uses 350 minutes in one month.

Lesson 11-6: Mixture Problems

Mixture problems usually involve "conservation laws." When you mix two items together, the total weight will be the sum of the weights of the separate items. If you mix 100 pounds of potatoes with 5 pounds of butter, you will have 105 pounds of mashed potatoes. We say that the weight of what is being mixed is *conserved*. Money is also conserved in mixture problems. If 100 pounds of potatoes cost $50 and 5 pounds of butter costs $15, then 105 pounds of mashed potatoes cost $65. We say that the money spent on the materials being mixed is *conserved*. These two conservation concepts play an integral role in creating the equations that need to be solved when working with mixture problems.

Example 1

KMM HealthFoods makes and sells a trail mix that includes mixed nuts and dried fruit. The mixed nuts cost $2 per pound and the dried fruit costs $1.50 per pound. Chris needs to make 50 pounds of trail mix that will cost $1.80 per pound. How many pounds of mixed nuts and how many pounds of dried fruit will he have to use?

Solution: Let m represent the amount of mixed nuts and d represent the amount of dried fruit. Because Chris needs to make 50 pounds of trail mix, so we have our first equation:

$m + d = 50$

Now we are ready to work with the money involved: the mixed nuts cost $2 per pound, so if we use m pounds of mixed nuts it will cost us $2m$. The dried fruit costs $1.50 per pound, so if we use d pounds of dried fruit it will cost us $1.5d$. Chris will end up with 50 pounds of trail mix that will cost $1.80 per pound, so the total cost is:

$$50 \text{ pounds} \cdot \frac{\$1.8}{\text{pound}} = \$90$$

We can generate our second equation:

$$2m + 1.5d = 90$$

The first term is the cost of the nuts, the second term is the cost of the fruit and the third term is the total cost of the trail mix. We have a system of 2 equations that we can solve:

$$\begin{cases} m + d = 50 \\ 2m + 1.5d = 90 \end{cases}$$

Use the substitution/elimination technique to solve this system of equations:

$$d = 50 - m$$

Use the first equation to solve for d: $2m + 1.5(50 - m) = 90$

Substitute $50 - m$ in for d: $2m + 75 - 1.5m = 90$

Distribute the 1.5: $0.5m + 75 = 90$

Simplify: $0.5m = 15$

Divide both sides by 0.5: $m = 30$

Chris needs to use 30 pounds of mixed nuts.

The amount of dried fruit is found by using the substitution equation to solve for d:

$$d = 50 - m$$
$$d = 50 - 30 = 20$$

So Chris needs to use 30 pounds of mixed nuts and 20 pounds of dried fruit in order to make the trail mix.

Lesson 11-6 Practice

1. Tim is running a successful landscaping business. If he hires 2 crew chiefs and 5 cutters, his daily payroll is $800. If he hires 3 crew chiefs and 8 cutters, his daily payroll is $1,240. How much does Tim pay his crew chiefs each day?

2. Dan sells fresh poultry to the local butchers. He sells 10 turkeys and 5 chickens for $120. He also sells 15 turkeys and 10 chickens for $190. How much does Dan charge for a chicken?

Answer Key
Lesson 11-2

1. Set up a proportion between the miles traveled on 12 gallons of gas to the gas mileage in general: $\frac{x}{12} = \frac{55}{1}$. Solve for x.
   The car can travel 660 miles on 12 gallons of gas.

2. Set up a proportion between the price for 44 squares and the price per square: $\frac{x}{44} = \frac{80}{1}$. Solve for x.
   Alan would earn $3,520 to shingle the roof.

3. Remember that percent means "out of 100" and a tax rate of 22% means $22 out of every $100 goes to the IRS. Set up a proportion between the tax on $30,000 and the tax on $100: $\frac{x}{30,000} = \frac{22}{100}$. Solve for x.
   Julia will owe $6,600 to the IRS.

Lesson 11-3

1. Set up a system of two equations and two unknowns:
   $$\begin{cases} x + y = 65 \\ x - y = 25 \end{cases}$$
   Solve the system of equations. The two integers are 45 and 20.

2. Set up a system of two equations and two unknowns:
   $$\begin{cases} x + y = 40 \\ x = 2y - 5 \end{cases}$$
   Solve the system of equations. The two integers are 25 and 15.

Lesson 11-4

1. Substitute into the equation $r \cdot t = d$.
   It would take Steven 5 hours to drive 300 miles.

2. Use the rate equation: $r \cdot t = d$. The distance is 35 miles and the time is 42 minutes, or $\frac{7}{10}$ of an hour. Solve for the rate.
   Cathy must travel at a rate of 50 miles per hour to get to work on time.

3. The total time that Ken will devote to his travels is 11 hours. The time he spends driving can be found using the rate equation $r \cdot t = d$. The driving distance is 300 miles and the rate is 60 miles per hour. The total driving time is 5 hours.
Ken will be able to spend 6 hours shopping at the mall.

4. Nicole will drive 44 miles at a rate of 48 miles per hour. Use the rate equation to find the time it will take Nicole to drive to the airport. The time is $\frac{11}{12}$ hours, or 55 minutes.
Nicole left at 5:10, so she will get to the airport at 6:05 P.M.

Lesson 11-5

1. Set up a system of two equations and two unknowns:
$$\begin{cases} 10t + 5f = 200 \\ t = 2f \end{cases}$$
Solve the system of equations.
Alic has 8 five-dollar bills and 16 ten-dollar bills.

2. The equation to model the cell phone bill is
$$bill = 40 + 0.35(t - 300)$$

Evaluate this function when $t = 350$.
The cell phone bill will be $57.50.

Lesson 11-6

1. Set up a system of two equations and two unknowns:
$$\begin{cases} 2c + 5u = 800 \\ 3c + 8u = 1,240 \end{cases}$$
Solve the system of equations.
Tim pays the crew chiefs $200 per day.

2. Set up a system of two equations and two unknowns:
$$\begin{cases} 10t + 5c = 120 \\ 15t + 10c = 190 \end{cases}$$
Solve the system of equations.
Dan charges $4 for a chicken.

Final Exam

1. Which of the following is a correct classification of the number 3.1415?

   a. Irrational

   b. Rational

   c. Natural

   d. Integer

   e. Whole

2. The least common multiple of 150 and 200 is the same as which of the following?

   a. The least common multiple of 125 and 225

   b. 50

   c. 3,000

   d. The least common multiple of 24 and 25

   e. None of the above.

3. Solve the inequality $5 < 1 - 3x \le 10$

   a. $\left(\dfrac{4}{3}, 3 \right]$

   b. $\left[-\dfrac{11}{3}, -\dfrac{4}{3} \right)$

   c. $\left[-3, -\dfrac{4}{3} \right)$

d. $\left(-\infty, -\dfrac{4}{3}\right) \cup [-3, \infty)$

e. None of the above.

4. Solve for x: $|3x - 6| = 9$

   a. $x = -1$

   b. $x = 1$

   c. $x = 1$ or $x = 5$

   d. $x = -1$ or $x = 5$

   e. None of the above.

5. Solve the inequality: $|x + 1| \geq 4$

   a. $(-5, 3)$

   b. $[-5, 3]$

   c. $(-\infty, -5] \cup [3, \infty)$

   d. $(-\infty, -5) \cup (3, \infty)$

   e. None of the above.

6. The Walker family will be attending a wedding. Jasper and Kelsey leave a day early and arrive in the town where the wedding will take place in 3.5 hours. Mary and Cindy leave the morning of the wedding, drive an average of 10 mph faster than their parents did the day before, and arrive in 3 hours. How far away was the town from the Walker residence?

   a. 165 miles

   b. 180 miles

   c. 60 miles

   d. 210 miles

   e. None of the above.

7. The sum of two consecutive positive integers whose product is 552 is:

   a. 23

   b. 45

   c. 54

   d. 47

   e. None of the above.

8. The midpoint of the line segment with endpoints (4,5) and (–6,–3) lies in:

   a. Quadrant I

   b. Quadrant II

   c. Quadrant III

   d. Quadrant IV

   e. None of the above.

9. Find the product of the solutions of the quadratic equation:
$2x^2 - 11x + 12 = 0$

   a. 4

   b. 6

   c. $\dfrac{121}{6}$

   d. $\dfrac{8}{3}$

   e. None of the above.

10. The distance between the x-intercept of the line $2x + y = 4$ and the y-intercept of the line $x + 4y = 10$ is:

   a. 6

   b. $\dfrac{\sqrt{29}}{2}$

c. $\dfrac{\sqrt{41}}{2}$

d. $\sqrt{116}$

e. None of the above.

11. The sales of various types of lawn and garden tools vary according to the season. At a certain hardware store, the monthly sales of shovels, S, declines from July to October, whereas the monthly sales of rakes, R, increases during this same interval. Suppose that the sales of these two items during the time period July to October can be modeled with the equations:

$S(t) = 64 - 6t$

$R(t) = 17t - 97$

where t is the month ($t = 7$ corresponds to July). In which month does the number of rakes sold equal the number of shovels sold?

a. July

b. August

c. September

d. October

e. None of the above.

12. The product of two consecutive whole numbers is 240. Find the smaller of the two numbers.

a. 14

b. 15

c. 16

d. 17

e. None of the above.

13. Find the discriminant of the quadratic polynomial: $4x^2 - 5x - 2$

    a. −7

    b. 57

    c. 42

    d. −22

    e. None of the above.

14. Find the slope of the line whose x–intercept is (2, 0) and whose y–intercept is (0, –3).

    a. $\dfrac{2}{3}$

    b. $-\dfrac{2}{3}$

    c. $\dfrac{3}{2}$

    d. $-\dfrac{3}{2}$

    e. None of the above.

15. If the perimeter of a rectangular flag is 34 inches and the diagonal is 13 inches, what is the area of the flag?

    a. 12 in²

    b. 60 in²

    c. 289 in²

    d. 169 in²

    e. None of the above.

16. Find the equation of the line passing through (3, 6) and (12, –6).

    a. $y = -\dfrac{4}{3}x + 10$

    b. $y = -\dfrac{3}{4}x + \dfrac{33}{4}$

c. $y = -\dfrac{4}{5}x + \dfrac{42}{5}$

d. $y = \dfrac{4}{3}x + 2$

e. None of the above.

17. What is the degree of the polynomial $-3x^2 + 6x^4 - 4x^5 + 7$?

   a. 2

   b. 4

   c. 5

   d. 7

   e. None of the above.

18. Find the sum of the zeros of the quadratic function:
    $x^2 - 20x + 64$

   a. 20

   b. 40

   c. 34

   d. 12

   e. None of the above.

19. The y-coordinate of the solution to the system of equations
    $\begin{cases} 2x - 4y = 11 \\ y - 3x = 2 \end{cases}$ is:

   a. Between –1 and –2

   b. Between 1 and 2

   c. Between –2 and –3

   d. Between –3 and –4

   e. None of the above.

20. A combination of 12 coins consisting of quarters and nickels is worth $1.60. How many quarters are there?

    a. 4

    b. 5

    c. 6

    d. 7

    e. None of the above.

21. Columbian coffee costs $6 per pound, and French Roast coffee costs $9 per pound. How many pounds of Columbian coffee should be mixed with French Roast coffee to obtain 100 pounds of a blend that costs $7.65 per pound?

    a. 35 pounds

    b. 45 pounds

    c. 55 pounds

    d. 65 pounds

    e. None of the above.

22. Craig paid $31.25 in cab fare from the airport to the hotel. The cab charged $3.75 for the first mile plus $2.50 for each additional mile. How many miles did the cab travel from the airport to the hotel?

    a. 11 miles

    b. 12 miles

    c. 13 miles

    d. 14 miles

    e. None of the above.

23. Subtract: $-2p+2w+7$ from $5p-4w+9$

    a. $-7p+2w-16$

    b. $7p-2w+16$

c. $-7p+6w-2$

d. $7p-6w+2$

e. None of the above.

24. Simplify: $\left(-4x^{-2}y\right)\left(-2x^5y^3\right)$

a. $\dfrac{8y^3}{x^{10}}$

b. $-8x^3y^4$

c. $-8x^{10}y^3$

d. $8x^3y^4$

e. None of the above.

25. Find the distance between the point (–1, –3) and (2, 3).

a. $3\sqrt{5}$

b. 1

c. $3\sqrt{3}$

d. 3

e. None of the above.

26. If $(x + 4)$ is a factor of $-x^2 - 11x - w$, then the value of w is:

a. –60

b. –28

c. 28

d. 60

e. None of the above.

27. One leg of a right triangle measures 15 inches, and the hypotenuse measures 17 inches. What is the perimeter of the triangle?

a. 32 inches

b. 40 inches

c.   60 inches

d.   127.5 inches

e.   None of the above.

28. Solve the inequality: $|3x+1| \leq -5$

   a. $x \leq -2$

   b. $x \geq -2$

   c.   All real numbers

   d.   No solution

   e.   None of the above.

29. At what point do the lines $y = 4x + 7$ and $y = -4x - 11$ intersect?

   a. $\left(-\dfrac{5}{4}, 2\right)$

   b. $\left(\dfrac{9}{4}, 16\right)$

   c. $\left(-\dfrac{9}{4}, -2\right)$

   d.   The lines do not intersect.

   e.   None of the above.

30. During a local performance of a play, the box office sold 247 tickets and collected $1,716. If a posh ticket cost $8 and a general admission ticket cost $6, compare the number of posh tickets to the number of general admission tickets sold.

   a.   13 more posh tickets than general admission tickets were sold.

   b.   25 more general admission tickets than posh tickets were sold.

   c.   13 more general admission tickets than posh tickets were sold.

   d.   25 more posh tickets than general admission tickets were sold.

   e.   None of the above.

Final Exam Answer Key

1. b	7. d	13. b	19. d	25. a
2. d	8. b	14. c	20. b	26. c
3. c	9. b	15. b	21. b	27. b
4. d	10. c	16. a	22. b	28. d
5. c	11. a	17. c	23. d	29. c
6. d	12. b	18. a	24. d	30. c

INDEX

I N D E X

INDEX

DENISE SZECSEI earned Bachelor of Science degrees in physics, chemistry, and mathematics from the University of Redlands, and she was greatly influenced by the educational environment cultivated through the Johnston Center for Integrative Studies. After graduating from the University of Redlands, she served as a technical instructor in the U.S. Navy. After completing her military service, she earned a PhD in mathematics from the Florida State University. She recently returned to graduate school to study epidemiology and biostatistics at the University of Iowa. She has been teaching since 1985, and hopes that the FSU Seminoles and the UI Hawkeyes never meet in a BCS bowl game.

ABOUT THE AUTHOR